Elizabete Rodrigues Sales

Digital Technologies

Elizabete Rodrigues Sales

Digital Technologies

Its place in teaching practices

ScienciaScripts

Imprint

Cover image: www.ingimage.com

This book is a translation from the original published under ISBN 978-613-9-61645-9.

Publisher:
Sciencia Scripts
is a trademark of
Dodo Books Indian Ocean Ltd. and OmniScriptum S.R.L publishing group

120 High Road, East Finchley, London, N2 9ED, United Kingdom
Str. Armeneasca 28/1, office 1, Chisinau MD-2012, Republic of Moldova, Europe
Printed at: see last page
ISBN: 978-620-7-88589-3

Table of contents:

ACKNOWLEDGMENTS

With regard to the completion of this work, I have a few special thanks to make.

To my beloved God, for all the good things that have happened to me since entering the Master's program. His omnipresence in my life is real. To Him be all the honor and glory.

To my dear parents, Francisco Sales and Maria de Lourdes Rodrigues Sales (*in memoriam*), for their love and educational role in my upbringing, as well as for the example of honesty and ethics they left their children.

To my beloved son, Rafael Rodrigues Sales Lazaroto (my Rafa), for his affection and understanding in all the moments of absence during the time I spent in the city of Sao Leopoldo (RS). From him came the inspiration that gave me the strength to continue in the most difficult moments of this journey.

To my dear siblings, Margarida, Margareth, Leonardo and Maria Francisca, my sister by upbringing and by heart. I am eternally grateful to them for their support at all times.

To the Ministry of Education (MEC), the Technological Secretariat of Education (SETEC) and the Federal Institute of Piauí (IFPI), through their managers, for the opportunity and investment in the qualification of teachers in the Federal Education Network.

"To the Master with affection", Prof. Dr. Telmo Adams, an advisor of extreme patience, knowledge and dedication to what he does. With his support, it was possible to carry out this work without stress. All my respect and affection to him.

To the professors Daniel de Queiroz Lopes and Carla Beatris Valentini, doctors on the dissertation's qualification and defense committees, who made this process a meaningful and enriching opportunity for me, a beginner in scientific research.

To the professors of the Postgraduate Program (PPG), Master's in Education at UNISINOS, for their welcome, enthusiasm and awareness of the important task of teaching. Likewise, to the staff of the PPG Secretariat, Library, Cafeteria and all the other sectors that make up the UNISINOS complex, for their welcome and professionalism.

In particular, to my dear brothers Auri, Schutz, Stanislaw, Afonso and Gabriel, residents of the Cristo Rei Spirituality Center (CECREI), for their welcome and the way in which they made this a wonderful and enriching experience. Happy memories that I will cherish forever.

I would also like to thank the CECREI administration team, led by Felipe, a responsible administrator and professional, without forgetting the human side of his relationships with his guests. I would also like to thank the reception, outpatient, secretarial, bookshop, cafeteria and cleaning teams for their warmth and welcome.

To the General Director, the Pedagogue and the teachers of the Municipal School of Teresina, which was the empirical field of my research, for welcoming me and being willing to cooperate, participating in the interviews, providing information that was fundamental to the production of this dissertation; as well as to the coordinator of the Teresina Technology Center (NTHE), for granting the interview, in addition to providing important documents on the implementation of the NTHE.

So, once again, I would like to thank God for having put special people in my life to share with me pleasant moments of great happiness and respect during our time together at CECREI, where we stayed for almost four months. First of all, I'm referring to Professor Madeira, a true friend, almost a brother, but already a brother in faith. With him I shared moments of great joy, others of sadness and anguish, but always sharing the feeling of hope, in the certainty that one day everything would pass and that victory was already determined by God.

I would also like to thank my dear friend Hugo Lenes Menezes for his availability and collaboration when I sought him out.

To my dear friend Ivan Oliveira, for his companionship during the time we were in Sao Leopoldo, always available when asked; as well as for his support in accompanying me during some of the interviews.

To Enói, a serious person with integrity, a great friend.

To Raíssa, Darlem and Nelymar, also participants in the master's program, for their companionship at CECREI.

To my colleague Sonia Matos Moutinho, for her help in formatting the "Ficha Catalográfica" of this dissertation.

I would also like to thank my other colleagues from the master's program for their company during class time and during some leisure time.

Finally, I would like to thank everyone who, directly or indirectly, contributed to this work.

One day you learn [...] You learn that no matter how badly your heart has been broken, the world won't stop for you to mend it.
Learn that time is not something you can turn back. So plant your garden and decorate your soul, instead of waiting for someone to bring you flowers. And you learn that you really can endure... that you really are strong, and that you can go much further after thinking you can't do it anymore. And that life really does have value and that you have value in the face of life! Our doubts are traitors and cause us to lose the good that we could achieve if it weren't for the fear of trying.
William Shakespeare

SUMMARY

The research looks at digital technologies (DT), involving the computer, information and communication technologies (ICT) and the *internet*, configured as mediating resources for teachers' pedagogical practices in the teaching process, and, from there, their implications and effectiveness in schools in contemporary times. The work is a qualitative study and is set up as a case study. It seeks to obtain from the subjects their conceptions of digital technologies and pedagogical practices. To this end, a study was carried out in a municipal public school in Piauí, whose teachers have at their disposal technical objects such as a *datashow*, a *notebook* and a computer laboratory acquired in 2002 through the National Educational Technology Program (ProInfo) and access to the *Internet* through the Broadband in Schools Program configured in a *WiFi* (*Wireless Fidelity*) technology system. The research subjects are the teachers at the school and were selected beforehand using a questionnaire. The methodological procedures, in addition to the interview in the empirical field with semi-structured questions, included collecting information during the research process, through observation in the empirical field, reading, access to computerized databases, *websites,* books, academic papers, articles, among others available on the *web*. The theoretical framework is based on the contributions of contemporary scholars such as Adams (2010), Castells (2004), Demo (2002, 2007), Freire (1987, 1996), Lévy (1999), Lopes (2010, 2011), Moran (2004), Santos (1994, 2006), Schlemmer (2011), Schwartz (2008), Tedesco (2004) and Valente (1999). The results of the research show that digital technologies are present in the pedagogical practices of teachers, most of whom use them in teaching activities. The reality of digital inclusion in pedagogical practices is mostly the result of teachers' efforts, whether in practice with individual experiences or through qualification by the NTHE. All the teachers are aware of the urgent need to appropriate these innovations in order to adapt to the new profile of the student who demonstrates skill in handling these digital technologies, which are present in everyday school life. At the Teaching Unit, the computer lab, which was set up to be a space for promoting digital inclusion, is not working properly, nor in accordance with the objectives proposed in Decree No. 6.300, of December 12, 2007, which deals with ProInfo.

Keywords: Digital Technology, Pedagogical Mediation, Digital Emancipation.

DESCRIBING THE SCENARIO IN FOCUS

This work is organized into six chapters. In the first chapter, entitled "Theme and focus of the research: context and presentation", I highlight the changes that the world is currently going through, given the new configurations in public policies, especially in education, as a result of digital technological advances, with an emphasis on the area of information and communication. From this, I can see the need to prepare people to live with an emancipatory attitude towards these new challenges. In this sense, I point out that this scenario permeates education in broader terms, requiring federal, state and municipal governments to adopt forms of democratic management that include actions for digital access and inclusion, as a right of citizenship.

I am describing the way in which digital technologies[1] are interfering in teaching practices and everyday school life. In this sense, the discussion I propose in this dissertation looks at the conditions in which digital technologies are used in pedagogical practices in the teaching process, how these digital resources are attributed in the conception of the teachers at the school under study, and how they can contribute to the emancipation of the subjects. Together with this problematization, I define the research objectives and present, throughout the text, the questions that guide the work.

In the second chapter, which I defined as "Identifying paths", I explained my background, education and teaching, with the aim of explaining why I chose to research education with an emphasis on digital technologies, as well as my relationship with these themes. I then try to identify the social relevance that this research can have for education, for the school community and for me as a researcher and teacher at a technological teaching institution.

The third chapter, entitled "Digital technologies: challenges and possibilities in education", briefly describes the advances in digital technologies (DT), particularly in the area of information and communication technologies (ICT) and the technological resources for these purposes. It highlights the changes in education and the Brazilian federal government's investments in programs that focus on digital inclusion. Here, I highlight the National Program for Information Technology in Education (ProInfo), describing historical data on ProInfo in Piauí, addressing the creation of the Teresina Technology Center (NTHE), its objectives and its actions in the municipality, as well as some data on the Broadband in Schools Program. All this information is directly related to my object of interest. These actions ensure that public schools are committed to their respective regions, creating technological solutions for sustainable development.

I also explain the conditions in which digital technologies are used by teachers in the teaching process, trying to relate the ways in which these digital technologies can contribute to the construction of knowledge, as well as to the formation of the digitally emancipated subject[2] . I am focusing on pedagogical mediation and digital emancipation, seeking support from authors who theorize on the subject. In this sense, the work will be enriched by dialogues with contemporary scholars of digital technology in education, such as Adams (2010), Demo (2007), Lévy (1999), Lopes (2010), Moran (2001), Oliveira (2001), Santos (1994, 2006), Schlemmer (2011), Schwartz (2007, 2008), Tedesco (2004), Valente (1999) and others.

In the fourth chapter, entitled "Pedagogical mediation and technologies in the teaching process", I try to evaluate the interaction between digital technologies and pedagogical mediation, considering them to be the guiding thread of the research.

Over time, education policies have undergone significant changes in the field of digital technologies. Federal public bodies are looking for strategic alternatives that can contribute to educational improvement. In this context, we are faced with the need for awareness, reflection and, above all, effective decision-making on aspects of digital exclusion and the process of including subjects in the field of digital technologies. I'm focusing on the period from the mid-20th century to

[1] Digital technologies, in this dissertation, correspond to computers and their various educational *software*, *datashows*, cell phones, *smartphones*, *internet* connections and the services they provide.

[2] By digitally emancipated subject, I adopt Schwartz's (2008) idea of a digital education that aims not only at the social inclusion of subjects through digital technologies, but mainly through the conscious, critical and autonomous use of these resources.

the present day, 2013, because I consider it to be a period that has seen a significant change in the teaching and learning process in everyday life in the classroom.

Here, I highlight the concept and evolution of the use of technical objects in order to gain a better understanding of what a teacher will value when planning and choosing teaching resources to work with educational content in the teaching process. In this respect, I would highlight what Santos (1994, p. 50) says about technical objects: "Pre-existing objects are aged by the appearance of more technically advanced objects, endowed with superior operational quality".

In the fifth chapter, entitled "Methodology", I explain the methodology used, based on reading Malheiros (2011), Minayo (2011) and Gil (2010), presenting theoretical foundations on the qualitative aspects of research and on the method configured as a case study. Next, I outline the section and the field in which the investigation will take place, as well as identifying the research subjects, procedures and techniques for data collection. Finally, I present a set of activities carried out since the beginning of the research: visits to the school, meetings with teachers, focused informal dialogues, observations and interviews, data which I transcribe in the field diary.

In the sixth chapter, entitled "The paths of analysis and results", I make a general analysis of the results collected in the empirical field, based on three thematic axes of analysis, in line with the problematization and objectives of the research. The axes include: teachers' experience with digital technologies in pedagogical mediation; the training process for the use of digital technologies in education; and teachers' conceptions of the computer lab (ProInfo).

Finally, in the section entitled "Final Reflections", I try to bring the experience to a close, bearing in mind that the subject is not exhausted in this investigation, as the future holds great surprises.

CHAPTER 1

RESEARCH THEME AND FOCUS: CONTEXT AND PRESENTATION

The world is currently undergoing major changes, given the new configurations in public policies, the economy and, above all, education, as a result of technological advances in the area of information and communication. It is therefore essential to prepare people to live with the new challenges of this scenario. These permeate education in the broadest terms, demanding democratic forms of management from federal, state and municipal governments, which include measures for digital access and inclusion, a citizen's right as opposed to the state's duty.

What's more, this reality demands appropriate management, in other words, an organizational model that meets the demands of 21st century education. This perspective is being assimilated by the Federal Government's policies through the main targets set for the country's education system, as part of a larger development project for Brazilian society. In this sense, Moran (2001) points out that:

> Society's political project must be to find ways of reducing the gap between those who can and those who can't afford access to information. Public schools and underprivileged communities must be guaranteed access if they are not to be condemned to definitive segregation, technological illiteracy and fifth-class education (MORAN, 2001, p. 51).

A document published by the Brazilian *Internet* Steering Committee[3] , following a survey on the use of information and communication technologies in Brazilian schools (2011, p. 21), states that:

> In Brazil, various government programs and actions encourage the use of ICT in education and demonstrate the commitment to invest resources for this purpose. However, there is still a lack of knowledge about the results of this use in relation to the achievement of the objectives and also about the achievement of the goals of educational programs, particularly with regard to the universalization of basic education and the improvement of teaching quality and school performance. Public policies in this field have focused on access to ICT and the development of infrastructure, but there has been little discussion of active participation, the development of skills, the digital literacy of students and also of teachers, pedagogical coordinators and principals. *Measuring and evaluating the impact of these technologies has become a necessity in the process of monitoring the construction and development of the information and knowledge society.* (emphasis added)

In order to meet this challenge, the Ministry of Education (MEC), through its public policies, has made available the National Program for Continuing Training in Educational Technology (ProInfo Integrated), a training program aimed at the didactic-pedagogical use of Information and Communication Technologies (ICT) in everyday school life; the National Educational Technology Program (ProInfo), which aims to train people to work with new computer and telecommunications technologies and prepare them to enter a new culture, supported by technology that supports and integrates interaction and communication processes (MEC, 2008); as well as making the One Computer per Student Program (ProUca) available to public basic education schools; the Teacher's Portal with educational *software* and the Broadband Program with *internet* access, among others. All of this is aimed at promoting the digital inclusion of teachers and students in the public school system, encouraging the creation and socialization of new ways of using digital technologies.

Schwartz (2008) evaluates the process of digital inclusion as actions that promote universal access to digital technologies and information and communication technologies, which contribute to reducing social differences, collaborating in the emergence of strategies for generating income and reducing unemployment. Focusing on the current federal government programs that prioritize digital inclusion, Schwartz highlights the main actions that involve this issue. For the author, the market, work and opportunities have now taken on a different configuration, represented by the offer of new

[3]The Brazilian *Internet* Steering Committee (CGI.br) was created by Interministerial Ordinance No. 147 of May 31, 1995, as amended by Presidential Decree No. 4829 of September 3, 2003, to coordinate and integrate all Internet service initiatives in the country, promoting the technical quality, innovation and dissemination of the services offered. More information at: http://www.governoeletronico.gov.br/o-gov.br/comite-gestor- de-internet. Accessed on: October 2013.

forms of work, through the intensive use of ICT; by the globalization of services and investments in training for citizenship, which are models of collective or shared access to the *Internet*; finally, by education in the information society: certification in information and communication technologies on a large scale (SCHWARTZ, 2008, p. 1). However, Lévy warns that it is not simply a question of:

> [...] to use technologies at any cost, but rather to consciously and deliberately accompany a change in civilization that profoundly questions the institutional forms, mentalities and culture of traditional educational systems and, above all, the roles of teacher and student (LÉVY, 2000, p. 172).

The reflections of Schwartz (2008) and Lévy (2000) highlight the importance of investing in digital inclusion programs as a means or path to social inclusion. In the certainty that digital technologies can contribute to social justice, this research admits as its main issue that the ability to access and acquire new knowledge through digital technologies is an analysis for social inclusion to occur. According to Demo (2007), the main issue involves "the socio-economic, political, socio-cultural, individual and social opportunities that are becoming increasingly conditioned by digital skills" (DEMO, 2007, p. 15).

The scope of the federal government's actions ensures that public schools are committed to their respective regions, creating technical and technological solutions for sustainable development, including both access and digital and social inclusion[4] . Similar investments are aimed at regional development through the ProInfo, ProInfo Integrado, ProUca and Broadband in Schools programs with Proposals, Projects and Adhesion defined after consultation with education managers and teachers from state and municipal schools.

The policies in question have been implemented through the Educational Technology Centers (NTE) and Municipal Technology Centers (NTM) in Brazilian states and municipalities, whose responsibility is to promote the training of teachers to manage the use of computers and Information and Communication Technologies (ICT) in the classroom, offering technical support and monitoring to support teachers who use the ProInfo laboratory at school.

The introduction of computers, the *internet* and other digital technological resources into the educational environments of public schools has required teachers to better understand how to use these devices as a pedagogical mediation strategy, with a view to keeping up with the technological developments that are dominant in today's world of formal work and increasingly widespread among students. This requires a transformation in the way education professionals teach, from being mere transmitters of information and knowledge to mediators of the teaching and learning process.

It can therefore be seen that the new configuration of educational environments is largely the result of digital technological development, which demands certain attributes from schools and their managers, such as: commitment, initiative and, even more so, a new look at education, which also implies articulation for access to Federal Government programs, incentives for teacher qualification, adaptation of environments that favors digital inclusion, technical-administrative support and adequate infrastructure for installation and permanence of equipment in working order that contributes to teaching performance. Reflecting on the issue, I ask myself: Does making computers and the internet available guarantee teaching success in the process of building knowledge and enhancing human knowledge? In this proposal, I adhere to an understanding of "human" as a being in continuous construction, which is organized historically in space/territory and in social interaction. This presupposes an education focused on the social relations of subjects in a process of continuous transformation, which Oliveira (2001) understands to be:

> (...) a concrete work of production and social reproduction of human existence, in the material and spiritual spheres, through which the actors in the pedagogical situation relate to each other and to the natural and social world (OLIVEIRA, 2001, p. 102).

For the aforementioned author, the most important part of education lies in human formation

---

[4] Sustainable development based on digital technologies is understood as the idea of sustainable processes and actions to meet the needs of present and future generations, through the inclusion of individuals and communities to form sustainable networks.

and involves "assimilation, construction and cultural production", a concept linked to the formation of subjects in a more complete sense, also encompassing the understanding of being and being in the world, as well as the interaction and socialization of information.

As the contemporary world is characterized by a knowledge and information society, evidenced by the breadth of the different forms of communication and the use of technologies, it is essential for people to participate in social and working relationships. In this respect, intense reflection is required in the field of the performance and execution of educational actions aimed at digital inclusion. From this point of view on Information and Communication Technologies (ICT), Costa (2007) states that they bring:

> the possibility of democratizing and universalizing information with great potential for reducing social exclusion, although paradoxically they have produced a new type of exclusion in undeveloped countries: digital exclusion. In Brazil, digital exclusion is a social and political problem, as it stems from the poor distribution of income in the country. In general terms, digital inclusion is understood as a form of support for citizens in the perspective of an Information Society, preferentially targeting the populations that have the worst socio-economic conditions, i.e. the lowest chances of appropriating the benefits brought by ICT.

Twenty-first century society demands that people continue to be better informed and able to use Information and Communication Technologies (ICT), as the lack of technical knowledge of how to use technological resources is creating a group of excluded people (LÉVY, 1999). This can be seen in the demands of the world of work, which tends to offer opportunities to professionals with a minimum of computer knowledge, sometimes involving specialized *software* and the use of social media, among other skills. These requirements lead to the exclusion of unprepared individuals. In this respect, I note that the Federal Government's educational actions and programs are being made available to public schools with the aim of including these individuals in the world of technology and, consequently, in the world of work.

With regard to information and communication technologies (ICT), one of the biggest changes in education is the different ways of creating, organizing and socializing content determined by multimedia, which involves digital technologies used to create, handle, record and research content.

In this respect, Tedesco (2004, p. 116) points out that, through connectivity, ICT generates new ways of teaching and learning, new ways of representing knowledge, creating and organizing content, as well as socializing information. Based on this new educational scenario, I ask: What are the attributes that determine the emancipatory digital inclusion of education professionals, considering access to digital technological resources configured as pedagogical mediation tools?

Another approach to the issue comes from the contemporary Moran (2001), who questions the role of the teacher and the changes taking place in educational environments involving digital technologies, since the latter interact not only in the relationship of space and time, but also in the process of interaction and communication between subjects.

Today, in addition to the classroom, computer labs, mobile technologies, cell phones, *smartphones*, *ipods, internet* systems, social networks, *e-mail, chats*, among other options, are available.

With similar reflections, Lima (2001) points out that technological resources have a great fascination for students, as they are accompanied by a multimedia explosion, programs that combine *games* and educational information, virtual encyclopedias and alternatives that provide a different way of accessing information and knowledge. In this sense, the teaching and learning process that makes use of multimedia tools and language favors the ability to assimilate knowledge, better prepares students for their professional training, and at the same time expands their opportunities for inclusion in the world of work. This view is in line with the propositions of Moran (2001), for whom, in terms of ICT:

> [...] the first step is to try to make it possible for teachers and students to have frequent and personalized access to new technologies, especially the *Internet*. It is essential to have connected classrooms, adequate research rooms and well-

equipped laboratories. Teachers and students need to be able to buy their own computers through public or private funding.

From this perspective, I recognize the importance today of digital technological resources, especially ICT, aimed at teaching that adapts to the new demands of the teacher's profile and pedagogical practice, given the continuous and comprehensive access offered by the aforementioned resources, attributes valued by Moran in his intellectual production and works on the subject.

In this study, I collected information on digital access and inclusion, knowing that this process manifests itself in relationships with others, within an approach that considers, in general terms, sociocultural, cognitive and affective subjects according to the historical-cultural perspective of Vygotsky. In this context, according to Rego (1994, p. 16): Vygotsky adapted different branches of knowledge into a common context that does not separate the subjects from the cultural situation in which they are involved. He accepted the interaction between the subject and the environment as a defining characteristic of human constitution.

With this understanding in mind, I can highlight the fact that opportunities, or the lack of them, are linked to the concept of social class, the relationship between subjects and also to the actions of those in power.

In this respect, the MEC/SETEC's concern with building guidelines that take into account subjects from basic education can be observed, given the National Curriculum Parameters (PCN), the context of Basic Education, which states in Article 26 that Law 9.394/96, of December 20, 1996, determines the construction of curricula, in Primary and Secondary Education, with:

> (...) a Common National Base, to be complemented, in each education system and school, by a diversified part, required by the regional and local characteristics of society, culture, economy and clientele.

In view of the above, in this investigation I tried to study a Municipal Public School in Piaui, located in the northern part of the city of Teresina, capital of the state of Piaui. The school adopted and implemented Federal Government programs in the school community: ProInfo, acquired in 2002, and Broadband in Schools, with the aim of providing access and digital inclusion for teachers and students. The legal procedures were mediated by the Teresina Technology Center (NTHE) and the Municipal Education Department (SEMEC).

Given this context, I was interested to see how digital technologies are being used by teachers in their daily pedagogical practices at the school in question and how teachers perceive these resources, considered by education to be mediators and enhancers of the teaching process.

From this point, I ask: What is the teachers' conception of digital technologies in education and under what conditions are these resources used in their pedagogical practices to mediate the teaching process? And what is their view of the computer lab purchased by ProInfo and made available for activities in the school?

Based on this problematization, my intention was to analyze the context of a municipal public school in Piauí, involving the use of digital technologies in the pedagogical practices of its teachers. As a general objective, I decided to gather information from the teachers in order to understand the conditions under which they use digital technologies, investigating the influences that these resources have on teaching action and the meaning that the teachers give to these digital resources in the teaching process. From this perspective, I set myself the specific objectives of analyzing the conditions of teachers' pedagogical practices in the use of the computer lab and the use of digital technologies in teaching, identifying whether these correspond to pedagogical mediations that enhance digital emancipation; distinguishing teachers' conceptions of digital technologies as important resources in the pedagogical mediations of the teaching process; identifying, in teachers' activities, to a lesser or greater degree, the inclusion of technologies as digital didactic-pedagogical resources.

In order to guide the work, more precisely, in order to find ways of directing the results of the research, I constructed some guiding questions that are set out here: How can digital technological resources contribute to the development and improvement of the teaching process? What attributes determine emancipatory digital inclusion, given access to digital technologies (DT)? What influence do digital technologies have on teaching and the execution of teaching activities?

1.1 Assumptions about technological resources in education

The hypothesis of this research is that there is a high level of appreciation of digital technologies in education at a national level. However, I would like to highlight the difficulties in terms of infrastructure and technical support for maintaining the digital technological resources made available in the school environment, as well as the process of training teachers to use digital technologies. I suspect that, in didactic-pedagogical activities, digital technological resources, in most cases, are not being used as pedagogical mediations, that is, as a "system of relationships that generate a set of flexible steps that take place in educational processes", as Adams (2010, p. 37) teaches, but only as means of transmitting information and content.

Common sense has shown how involved and "enchanted" people are with digital technologies, especially information and communication technologies, all of which are used as educational technological resources and made available in different spaces and contexts. In this respect, it is worth mentioning Castell's (2007) statement:

> New communication technologies have created new spaces for knowledge. Now, in addition to schools, businesses, homes and social spaces have also become educational. Every day, more people study at home, due to the ease of access to information from thousands of databases, spread geographically around the world (CASTELL, 2007, p. 5).

On this subject, it is possible to state that information and communication technologies contribute to the construction of an increasingly globalized world, where people are encouraged to participate in this great universal network, innovating and creating new forms of interaction and communication between subjects, acquiring knowledge, socializing ideas, all mediated by digital technologies.

I note that today it seems imperative to use these technological resources in education. With regard to this hegemony, I can highlight the investments made by the Federal Government's actions, through the Ministry of Education, aimed at large-scale digital access and inclusion for teachers and students in Brazilian public schools, which are included in its Public Policies and Social Programs, such as the Broadband in Schools program and the Information Technology in Education Program (ProInfo), which in this investigation are part of my object of interest.

Although it is not the focus of my study, Distance Education (DE), which was created in 1995 during the government of President Fernando Henrique Cardoso, with the slogan "Democratization of Access", has today become an important part of the Brazilian education system, as Schlemmer (2011) observes:

> Since the Law of Guidelines and Bases (LDB) - Law No. 9.394/9624 recognized distance education, there has been a significant growth in the provision of training and qualification in this modality every year. The EAD. BR 2009 Census, carried out by ABED (Brazilian Distance Education Association), recorded a total of 1,075,272 students enrolled in distance learning courses at accredited institutions. The figures show that this type of teaching is growing by around 65% a year.

The actions of the Federal Government through the MEC, with regard to Distance Education, demonstrate new paths for education, aiming at the training of subjects and meeting the needs stimulated by advances in digital technologies.

In addition, at the educational level, the PCN's promote the use of digital technologies in the teaching and learning process, providing opportunities and creating educational perspectives through distance learning and other educational technologies.

1.2 Identifying the issue in Brazilian research and official documents

In relation to the area of interest "education and digital technologies", national literature has been emerging from within academia, through articles, master's dissertations and doctoral theses. As an example, I would mention Campos (2011), who, in his master's thesis, is concerned with raising questions about the formation of subjects in the school environment and the use of computers and the *internet*. In this sense, the author aims to find out if the teacher, when using technological resources, is able to articulate instrumental and pedagogical use, while asking: "After all, what changes with the arrival of these educational technologies and what influences do they have on teacher training, physical space and pedagogical conceptions?" (CAMPOS, 2011, p. 18),

Gengnagel and Nicolodi, in their study on "Public connection policies: challenges of the 21st century", discuss the importance and need for public policies in the area of education and, in particular, in the area of information and communication technologies. They analyze these educational policies and verify the potential of current federal government programs: the One Computer per Student Program and the National Broadband Program. They highlight the great importance of government guidelines that bring education into line with the trends seen and experienced around the world, especially with regard to technology.

Alves (1998), in his master's thesis, discusses new cognitive cartographies and analyzes the use of intellectual technologies by public school children in Salvador, Bahia. She points out that the history of humanity has been marked by the presence of intellectual technologies[5] , reorganizing the social context, mediating new ways of thinking and constructing knowledge. According to Alves, technologies have been re-signified, making it possible to redefine concepts. These technological elements are present in people's daily lives. However, we are faced with the formation of a new culture, a new way of thinking, requiring the school to rethink its pedagogical practice, especially with regard to the cognitive development of individuals.

Pacheco (2011), in his master's thesis, seeks to understand the process of digital education in everyday school life and reflects on the pedagogical practices of teachers, as well as addressing the re-signification of these practices with the possibility of building an inclusive school. However, the results of her research reveal that even with the promising advances in the field of training, the time is still far off to consider that digital education actually takes place in every school context.

In addition to these dissertations, I tried to direct my reading, in a more general sense, to texts on digital technologies and, in a more specific sense, to official documents from the Federal Government regarding public policy programs for digital inclusion, as well as works that refer to the pedagogical use of information and communication technologies; the qualification of teachers for the use of computers, the *Internet* and information and communication technologies.

The results of these researchers highlight the fact that it is imperative to invest in the qualification of teachers for the new educational scenario, which requires, at the very least, technical knowledge of how to use digital technological resources and, in particular, knowledge of information and communication technology. These academic works encourage reflection on similar resources in the educational process.

On the subject in question, the authors we have read are similar when they refer to the need for the following items in order for education to be of superior quality: teacher qualification through training courses in digital technologies; the representation of teachers and students in the current technological scenario; the school's responsibility in the learning process, focusing on the introduction of new technologies for digital access and inclusion; interdisciplinarity in the context of teaching and learning; interactivity and exchange relationships in the construction of knowledge. In terms of these new developments, the *internet* and computer networks are also highlighted, as they are considered to be benchmarks in the exploration of technological educational potential.

The theme proposed for my research, digital technologies: their place in pedagogical practices in a municipal public school in Piauí, with an emphasis on the federal government's initiative aimed at digital inclusion through ProInfo, is a different focus from other research in this area. Started in 1997, ProInfo had its pilot project distributed in five Brazilian states and in public basic education schools.

It is extremely important to constantly invest in studies aimed at making the most of new technologies within these programs. Teachers' actions aimed at using information and communication technologies require them to understand the educational potential of these resources in the process of digital emancipation.

In the state of Piauí, there is still a lack of research demonstrating the knowledge resulting from studies in the area of digital access and inclusion in state and municipal public schools, after they joined the ProInfo and Broadband in Schools programs.

5Alves identifies intellectual technologies as games, *software*, the *internet,* TV and video. Intellectual technologies are elements that promote the restructuring of human thinking, giving rise to new cognitive cartographies.

CHAPTER 2

IDENTIFYING PATHS

In this topic, I'll explain my academic career, training and teaching, with the aim of explaining why I chose this master's degree subject in Education, in the Education, Development and Technology line of research, of the Postgraduate Program in Education at the University of Vale do Rio dos Sinos (UNISINOS-RS), through an inter-institutional agreement (MINTER) between the latter and the Federal Institute of Education, Science and Technology of Piauí (IFPI), my institution of origin.

I'll start by saying that from a very early age, at the beginning of my adulthood, I had an artistic streak, more precisely in the area of visual arts, a characteristic that is repeated in some members of my family, as well as my admiration for the teaching profession.

In the 1970s, while I was still very young, I managed to get a job teaching Art Education in public schools in Teresina (PI). Despite the fact that I didn't yet have a teaching degree and was still in my first year of high school, I enthusiastically began teaching at four of the schools in the Zona Sul School Complex.

My academic background is in the technical field, as I have a degree in Social Communication with a major in Advertising from UNISINOS. But, already thinking of following the path of teaching, in 1990 I took a specialization course in Higher Education Teaching at the TUIUTI College in Curitiba (PR).

In 2000, in Teresina, I was approved through a simplified selection process to teach the Technical Advertising course at the then Federal Technological Education Center of Piauí (CEFET-PI). At that time, I started working on a social project with the students on the course, creating and producing an annual advertising campaign for the children with cancer assisted by the Piauí Women's Cancer Network (RFCC-PI). Already at that stage, I had my first experiences with the use of computers and computer graphics *software*, used as didactic-pedagogical resources in the computer lab. In 2004, I transcribed these activities, especially for their initiative and consistency, in the form of an experience report, which was then awarded a publication in the *Cadernos Temáticos* magazine published by the Ministry of Education (MEC) and the Secretariat of Technological Education (SETEC).

The previous year (2003), through a public examination, I began working as a teacher of basic, technical and technological education at the then CEFET-PI, now as a permanent teacher at the institution.

In 2008, CEFET-PI became the Federal Institute of Education, Science and Technology of Piauí (IFPI), with autonomy and university *status*. At the institution, I worked for some time as a Communications Officer, creating and producing various types of printed media to publicize institutional and academic activities. This task of creating and producing graphic pieces also required the use of computers, the *internet* and other digital technological resources.

As far as teaching is concerned, in the administration of the Visual Communication subject on the Visual Arts Technical course at IFPI, I develop the concepts and fundamentals of the graphic area. I cover all the theory surrounding the psychodynamics of colors and balance: symmetry and asymmetry, the history of typology (fonts), among other topics. In my teaching experience, I realize that the use of digital technologies as a working tool, especially the *internet* as a resource for searching for relevant content, as well as computer screen display equipment, computer graphics *software*, all of this enriches the teacher's knowledge and contributes to motivating the student in the learning process. However, I mustn't forget that in the teaching and learning process, the knowledge I have acquired throughout my academic and professional life is a resource that I have always used to enhance the dynamism of the classes I teach.

In my particular case, in view of the aesthetic-visual language displayed on screen exhibitors, I recognize that digital[6] media can favor and enhance, but not replace, content mastery in teaching performance. This is because the tasks of researching, preparing lessons, preparing class activities, among others, require the teacher to have the skills and knowledge to transform all this available

---

[6] A collective term used to refer to the means of communication (press, television, radio, *internet,* cell phones, etc.).

technology into enriching didactic-pedagogical resources, allied to daily performance.

The world of digital technologies offers infinite possibilities. I believe that, for the teacher, mastery of content and praxis in the area in which they work, especially concepts and values for the formation of citizens, become attributes of great value in the teaching process when mediated by appropriate resources.

I see that information and communication technologies, the computer and the *internet,* mobile technologies, cell phones, *smartphones*, *iphones,* among others, are resources that can contribute, if used properly, to improving activities that involve the construction of knowledge in the process of training subjects in educational institutions. A new scenario is emerging in formal education environments, and this change in the field of digital technologies needs to be recorded.

I believe that, in the educational process, good computer equipment, mobile or fixed technologies do not turn an education professional into a good teacher, although, nowadays, a good teacher can do much more if they have, in addition to mastery of content, technical knowledge for manipulating and using technical objects, creativity, motivation and computer technology in appropriate contexts.

Reviewing all these facts, I realize how much digital technologies, especially information and communication technologies, are present in my life and how much I use them professionally. In this context, I would point out that, on a personal level, these resources are also important for my relationships with friends and colleagues at work, when I have a computer, cell phone, *internet* and social media.

2.1 Social and academic relevance

Currently, the so-called "globalization" promotes access to information, the construction and socialization of knowledge on a large scale, all through technological resources, especially information and communication resources, which are increasingly present in people's daily lives. These are computers, cell phones, *smartphones,* iphones, *laptops*, *ipods, tablets*, among other resources. All of them, mediated by the *internet,* promote social, work, political and economic relations, transforming the world stage.

According to Schwartz (2007), the effects of the federal government's various actions to promote collective access to the *internet* and basic training in the use of technological tools, such as computers, are an open question. For the author, once the equipment has been installed, the initiative seems to run out of steam, and there are no indicators of the results of these investments. Against this backdrop, this research aims to gather data that could help to support this lack of information on the federal government's investments in digital access and inclusion policies.

Corroborating what Schwartz (2007) says, Oliveira (2001) suggests the need to recommend "information technology as an object of study and not just as a teaching and learning resource". According to the author, there is a need for research in this area that points to:

> The question of the use of information technology in education itself, based on experience and practices not developed by the *a priori* defense that this use is related to improving the teaching-learning process and meaningful learning; the culture of information technology, highlighting the properties of information resources, not reduced to the question of computer rationality and its analogy with human rationality. It is important to research, for example, the characteristics of computer language, not just in terms of its logical aspects, but above all in terms of its interaction aspects, which define different tribes of users; computer culture and its relationship with school culture and other cultural universes. It is worth asking: to what extent does the use of the *internet,* for example, favor the construction of an intercultural perspective at school or the strengthening of monocultural attitudes or prejudices against the culture of those who are different, or to what extent does the use of the *internet* imply a different culture, at the crossroads of cultures at school? (OLIVEIRA, 2001, p. 106).

This research paper is the result of my reflections and questions about the whole process of including technological resources in the educational system, more specifically in teaching practices, from which I have tried, through reading, to find a basis in authors who deal with the subject. In this regard, as can be seen, the aforementioned authors reinforce the need to investigate the results of investments and the use of information technology and the *Internet in* order to measure their effects on the process of building an expectation between cultures at school, aspects of interaction and identification of different users of digital technologies. In this context, I believe that the results obtained in this research can help managers, teachers and students to take a critical look at the performance of their activities when mediated by digital technologies.

To this end, I believe it is important to carry out ongoing evaluations to assess the results of the programs implemented and also to ascertain teachers' conceptions of the use of digital technological resources in teaching practices, which is the subject of my interest, as well as to gather data on the impact of technology on the educational process and on improving the quality of teaching actions, while also observing the process of building these realities.

Specifically, through this research paper, I study and gather information about the educational reality of a Municipal Public School in Piauí that has joined the Federal Government's PROINFO Program (MEC-SED), which encourages:

> [...] an evaluation culture in the different Brazilian states, through the participation of those involved: professors from Higher Education Institutions (HEIs), participants from the state and municipal Technology Centers (NTE and NTM), educators and students from the schools benefiting from the Program (CAPPELLETTI, 2012).

The relevance of my research into digital technologies and ProInfo in Piauí, in a municipal public school, lies in the fact that I was able to record the experiences, conceptions and also the teachers' involvement and interrelationship with digital resources, including the computer lab made available in the school through the program, and the *internet* access system acquired through the Broadband in Schools program, information that is open to analysis.

In this context, I would like to highlight the challenge that education professionals face in working with digital technologies in everyday school life, depending on investments from federal, state and municipal governments, the infrastructure conditions of schools, technical support, training and the time available for this, among other important factors. All of this interferes with the appropriate use, or not, of these technological resources during teachers' pedagogical practices.

CHAPTER 3

DIGITAL TECHNOLOGIES: CHALLENGES AND POSSIBILITIES IN EDUCATION

In Brazil, the introduction of information technology in education took place at the beginning of the 1970s, when the policy of computerization of the productive sectors adopted by the Brazilian government demanded autonomous scientific and technological training, requiring investments in the educational area so that it could provide support for the intended large-scale computerization. At the end of the 1970s, there was strong interest from the productive, economic and social sectors in access to computerization. In order to guarantee National Sovereignty, Law No. 7.232 was passed, which defined a market reserve for computer-related equipment industries, with the aim of developing national autonomy in science and technology (OLIVEIRA, 2006).

However, it was in 1979 that the computerization of Brazilian society began to be effectively programmed. The government started a structuring process and academia began to establish a new space for this issue. Changes occurred, but information technology always remained linked to sectors more focused on information processing and control (MORAES, 1997).

3.1 Digital technologies in an educational context

Today, in Brazil, the use of digital technology is a reality in schools and, notably, in the world outside them, as evidenced by the large number of computers, *laptops*, *tablets,* cell phones, *smartphones* and the like in use by the Brazilian population. This fact drives us to seek an understanding of the appropriate use of digital technologies in educational activities and to turn them into a reference point for integrating the school into the world of scientific production, culture, social relations and work.

However, having access to digital technologies such as computers and the *internet* is not enough. It is necessary to make choices in the search for information and the use of the most appropriate tools, allowing teachers to make changes in the planning and organization of content for the development of their didactic-pedagogical activities. In this sense, it is essential to raise awareness of the need for changes in teaching practices, with attention to solutions to the difficulties of everyday life at school and with a careful eye on innovations in the contemporary world scenario.

I reiterate that this change in the educational scenario begins with the understanding of teachers and managers in the performance of activities that organize, motivate and provide conditions for the development of teaching. In order to do this, it is necessary to adapt to the technological reality that notably permeates the physical educational environment, the training of education professionals, the organization of activities that promote access and digital inclusion, generating significant changes in the teaching and learning process.

Digital technologies and social media currently have a major influence on the daily lives of people in the school environment. The big challenge is to understand how computers, the *Internet* and other technologies can help to make teaching practices more attractive and effective. According to Moran (1997, p. 20): "The *Internet* alone does not change the process of teaching and learning, but the basic personal and institutional attitude towards life, the world, oneself and others." According to Demo (2001),

> [...] The days of reproductive classes are numbered, because not only do they rob students of the possibility of reconstructive learning, they also imbecilize them. An important part of learning is knowing how to handle, search for and produce information, so that we are not just manipulated objects. *An interesting lesson is one that leads to this, not one that hinders it.* (emphasis added) (DEMO, 2001 p.26, 27)

In this context, I recognize that teachers need to have mastery not only of a specific technique or content, but above all knowledge of the appropriate ways of teaching for an emancipatory education. Education for the new times is not expected to be simply an introduction to digital technologies that ensures professionalization, but an action that lasts throughout the lives of individuals. Preparing people to be successful in today's world is not just about imparting knowledge, but, above all, giving them the conditions and encouragement to develop skills and abilities that will enable them to act in their environment, finding knowledge for their lives and their relationships with the world.

The language of the media, full of images, movements and sounds, attracts the younger generations, which is why changes need to take place in schools and, in order for them to happen, educational institutions need to rethink, reorganize and reposition their own structure and curriculum. Innovative proposals must emerge, focusing on the training or continued training of teachers, in order to generate reflections on emerging themes and paradigms in education for digital inclusion. In this sense, I would highlight what Demo (2001, p.
29) calls reconstructive personal effort. For him, reconstructive learning is not about being a mere spectator or listener to someone transmitting information, but about reconstructing knowledge through an accompanied constructive process.

With reference to the educational reality in schools, I would highlight the inclusion of information and communication technology and its advances, to which it is possible to have access, a fact that keeps schools up to date through the use of digital technologies and technically advanced objects. In the teaching and learning process, the most important thing is to research and reflect on information.

At the present time, there is no denying that computers, mobile technology and the *Internet* are effective resources in the teaching and learning process. However, it is up to the teacher to make use of these resources coherently, always aiming to enhance the transmission and exchange of knowledge, in order to better adapt the contents of the subjects that make up the curricular matrix of the courses and the interrelationship of the subjects in the school environment. For this to happen, teachers need to be interested, knowledgeable and adapted to their realities, adapting digital resources to their didactic-pedagogical practices.

In this environment, information and communication technologies in the process of interdisciplinarity are, if well managed, resources that enhance the teaching and learning process, offering several benefits when compared to traditional teaching paradigms, such as offering the dissemination and instant sharing of information, the adaptation of activities that favor the personal characteristics of the subjects and the "intercrossing of cultures at school" (SCHWARTZ, 2007).

In this sense, it is of fundamental importance to provide students with new experiences, awakening their motivation, interaction in class and the socialization of educational content, developing a taste for reading, which culminates in the acquisition of knowledge and new learning experiences.

Castells (2004), Demo (2001), Levy (1997, 2012), Moran (2001) and Tedesco (2004) are authors who have made outstanding contributions to the field of education and digital technologies. They emphasize the fact that in the networked society, each individual is an agent for disseminating information and represents a point in the broad web of information.

For Castells (1999), quoted by Marques (2009), science and technology play an important role in these accelerated changes, influencing and being influenced by society. What characterizes the current technological revolution is not the centrality of knowledge and information, but the application of this knowledge and information to generate new knowledge and information communication processing devices, in a cycle of cumulative feedback between innovation and its use (MARQUES, 2009, p. 28).

The so-called network society shortens distances and time, bringing together people with similar interests and objectives. From this perspective, Levy (1997) conceptualizes what he calls cyberspace as a wide space where digital information travels in open communication through the worldwide interconnection of computers and computer memories.

In this global system, a large number of technological instruments are connected in cyberspace, such as videoconferences, social networks, among others, which are part of the multimedia universe, made up of sound, images, text and animation, and whose main objective is the socialization of information and the construction of knowledge. The new knowledge is integrated with methods of using and archiving digital information. Thus, it is understood that computers, mobile technologies and the *Internet,* as information and communication tools, are becoming technologies that are increasingly used and appropriated by individuals, and are entering different social strata every day.

The school, in its pedagogical practice, must bring students closer to this new reality through its educational policies and actions. ProInfo and Broadband in Schools are programs that have come to make it possible for schools to be included in the digital world as actions that give access to the use of computers, their applications and multiple resources, in short, to cyberspace. All of this, if combined with a political pedagogical project that is suited to educational changes, becomes an instrument that enhances teaching activities in the teaching process of public schools.

Moran defends these assumptions by stating that:

> We need to educate for democratic, more progressive and participatory uses of technology that facilitate the development of individuals. When children arrive at school, the fundamental learning processes are already significantly developed. There is also an urgent need for media education, in order to understand, criticize and use them as comprehensively as possible (MORAN, 2000, p. 50).

The Federal Government's initiatives for Digital Inclusion, through the Ministry of Education, help to train people for the challenges of the "globalization" of the markets, the scientific and technological society, the construction of knowledge and the socialization of information. According to Lopes and Schlemmer (2011):

> The idea of digital inclusion within the scope of public policies is in line with what is perceived in relation to the current national and global socio-cultural, scientific and technological scenario, which indicates an increasingly intensive use of DM in the construction and dissemination of knowledge. In this sense, appropriating DT has become a concern so as not to produce or intensify a new form of exclusion.

It is therefore important to collect data that can provide an evaluation of government investments. These are initiatives by means of research that presents the dynamics of educational management in the different federal programs and actions that aim to promote digital inclusion in Brazilian public schools, as well as verifying how these programs are being managed, such as Broadband in Schools, which installs an *internet* access system, ProInfo, which aims to set up computer labs in schools. Initially, 167,000 computers were distributed to 14,500 schools throughout Brazil (an average of ten machines per laboratory in secondary schools) (BRASIL, 2007).

I emphasize that the impact of technological advances in the 21st century is a reality that we cannot escape. In our daily lives, the intense presence of social media, especially information and communication technologies, and advances in mobile digital technology, allows us to broaden our world view. These facts provide new knowledge and new ways of apprehending and understanding reality.

Today, there is an urgent need to include these technologies in the school curriculum, thus preparing students to interact properly in this technological and computerized society. The need has arisen for a new educator, with a new profile, involving the role of mediator in teaching, who sees himself as an investigator of the reality of the students. Today, therefore, I believe it is imperative for teachers to be aware of all these demands arising from the educational scenario, which requires a new attitude from them. The digital technological environment demands changes in education and a re-dimensioning of teaching methodologies, capable of establishing horizontal relationships between educator and student, acting together as apprentices in the construction of knowledge. The adoption of digital technologies in classroom activities or educational environments can boost student learning, encouraging them to interact with the various means of digital communication, enhancing their creativity and providing them with digital emancipation.

The spread of multimedia computers and the expansion of electronic network connections in companies and families in the mid-1990s intensified the discourse on the need to incorporate such equipment into teaching work in education. Thus, by the end of the 1990s, practically all European Union countries, a large number of Latin American countries and several Eastern European countries had joined the formulation of government programs aimed at equipping their schools with computers and, above all, connecting them to the electronic network. In Brazil, it was no different: in 1997, the National Program for Information Technology in Education (ProInfo) was created, linked to the Ministry of Education's Secretariat for Distance Education (SEED), a proposal by the FHC government (1994-2002) to introduce these technologies into the public education system.

In its justification and objectives, the Program (1997) pointed out the need to bring school culture closer to the main technological advances in contemporary society and highlighted the educational possibilities of technical networks for storing, transforming, producing and transmitting information. The document emphasizes the democratization that computers can bring to students graduating from public schools, as they are placed on an equal footing with students from private schools. Within this proposal, "technological literacy" is considered essential, "just as important as knowing how to read, write and do math" (BRASIL/MEC/SEED/ProInfo, 1997, p. 4).

Today, the federal government's efforts in the field of digital technologies can be seen in education in Brazil. In turn, the federal government is working to promote digital inclusion in schools, encouraging and disseminating the use and supply of information and communication technology goods and services, through programs and Educational Technology Centers (NTE) set up in the Brazilian states. These develop digital training for teachers and education professionals. In this context, the evolution of educational processes mediated by the role of digital technologies is essential, as these processes enhance the socialization of subjects. In this direction, it is possible that the insertion of digital technologies into educational situations will produce new meanings, also contributing to the formation of digitally emancipated subjects.

From the point of view of an innovative perspective, digital technologies, more precisely information and communication technologies, are not just the introduction of an auxiliary resource for teaching, but also possibilities, mediation and the promotion of horizontally emancipatory socialization within digital culture. However, what is needed to address this reality are government initiatives that provide financial, political and structural support for the real implementation of projects in public schools throughout Brazil.

3.2 Digital inclusion programs in Brazil

In view of the complex contemporary world, its dynamics in the transformation of social relations, the economy and, consequently, the process of education on a global level, especially with regard to public policies in the area of education, I would like to highlight those linked to the area of information and communication technologies. In this sense, it is important to highlight the Brazilian programs that encourage people to access and include digital technologies in schools.

In Brazil, the development of educational information technology in the public sector began in the 1980s, when the Ministry of Education (MEC) set out to invest in projects through schools to computerize Brazilian society. In this context, the MEC created the conditions for the encouragement and development of projects involving the use of digital technologies in Brazilian society.

Brito (2006), cited by Marques (2009, p. 33-35), reports on important actions in Brazilian educational policy that record the history of public policies and their investment in programs for digital access and inclusion.

TABLE 1: Historical timeline of public policies on educational information technology in Brazil

YEAR	ACTION
1979	The Special Secretariat for Information Technology (SEI) has made a proposal for the educational, agricultural, health and industrial sectors, with a view to making computer resources viable for their activities.

1980	The Special Secretariat for Information Technology has set up a Special Education Commission to gather input in order to generate standards and guidelines for the area of information technology in education.
1981	I National Seminar on Informatics in Education (SEI, MEC, CNPq) - Brasilia. **Recommendations**: educational information technology activities should be guided by the cultural, socio-political and pedagogical values of the Brazilian reality; the technical and economic aspects should be balanced not according to market pressures, but according to socio-educational benefits; the use of computer resources should not be considered a new panacea for tackling educational problems; pilot projects of an experimental nature should be created with limited implementation, with the aim of carrying out research into the use of information technology in the educational process.
1982	II National Seminar on Educational Informatics (Salvador), with the participation of researchers from the fields of education, sociology, informatics and psychology. **Recommendations**: The study centers should be linked to universities, with an interdisciplinary character, giving priority to 2nd grade education, while involving other teaching groups. Computers should function as an auxiliary means in the educational process and should be subject to the aims of education and not determine them, their use should not be restricted to any teaching area; priority should be given to teacher training in terms of theoretical aspects, participation in research and experimentation, as well as involvement with computer technology and, finally, the technology to be used should be of national origin.

1983	**Creation of CEIE** - Special Commission for Information Technology in Education, linked to SEI, CSN and the Presidency of the Republic. This commission included members from MEC, SEI, CNPq, Finep and Embratel, whose mission was to develop discussions and implement measures to bring computers to Brazilian public schools. **Creation of the Educom project** - Education with Computers. This was the first official and concrete action to bring computers into public schools. Five pilot centers were created, responsible for developing research and disseminating the use of computers in the teaching-learning process.
1984	Officialization of the Educom project's study centers, which consisted of the following institutions: UFPE (Federal University of Pernambuco), UFRJ (Federal University of Rio de Janeiro), UFMG (Federal University of Minas Gerais), UFRGS (Federal University of Rio Grande do Sul) and Unicamp (State University of Campinas). The funding for this project came from FINEP, Funtevé and CNPq.
1986/87	Creation of the Advisory Committee on Information Technology for 1st and 2nd Grade Education (CAIE/SEPS) under MEC, with the aim of defining the direction of national information technology policy based on the Educom Project. Their main actions were: holding national competitions for educational *software*; drafting a policy document that they defined; setting up Educational Informatics Centres (CIEs) to serve around 100,000 users, in agreement with the National and Municipal Education Secretariats; defining and organizing training courses for teachers at the CIEs; and evaluating and reorienting the Educom Project.

1987	Elaboration of the Immediate Action Program in Informatics in Education, which had as one of its main actions the creation of two projects: Project FORMAR, which aimed to train human resources, and Project CIED, which aimed to set up Informatics and Education Centers. In addition to these two actions, the needs of education systems related to information technology in 1st and 2nd grade education were surveyed, the Educational Information Technology Policy for the period 1987 to 1989 was drawn up and, finally, the production of educational *software was* stimulated. The CIED Project was developed along three lines: CIES - Centres for Information Technology in Higher Education, CIED - Centres for Information Technology in 1st and 2nd Grade and Special Education; CIET - Centres for Information Technology in Technical Education.
1997 to 2006	The creation of PROINFO, a project that aimed to form NTEs (Nuclei of Educational Technologies) in every state in the country. These NTEs will be staffed by teachers who will have to undergo post-graduate training in educational information technology. There are currently several state and municipal IT in education projects linked to PROINFO/SEED/MEC. The UCA Project (One Computer per Student) is a federal government initiative which, since 2005, has been investigating the possibility of using *laptops* in schools.

Source: BRITO, G.S. (2008).

3.2.1 National Program for Information Technology in Education (ProInfo)[7]

ProInfo is an initiative of the Ministry of Education (MEC) through the Distance Education Secretariat (SEED). It was created on April 9, 1997, under number 522, in which the MEC and the National Council of State Education Secretaries (CONSED) established the Program's guidelines.

The project aims to invest resources in the training of teachers and technicians, with the aim of using digital technologies to support and[9] education in public primary and secondary schools in Brazil.

Adhering to this federal government policy, the Brazilian states, through their state and municipal education departments, have set up Educational Technology Centers (NTE). These are basic structures for the process of digital inclusion in schools, which work by advising on the planning and implementation of computer labs in state and municipal schools. They are also responsible for training teachers and technicians in the technical use of digital technologies as a didactic-pedagogical resource for the teaching and learning process.

Of the 450 Educational Technology Centers (NTE) and Municipal Technology Centers (NTM) set up in Brazil, 96 are in the Northeast. Of these, 13 are in the state of Piauí, including three municipal centers, including the Teresina Technology Center (NTHE), as shown in the table28.

[9] More information can be found at http://www.inclusaodigital.gov.br/outros-programas.

TABLE 2 - Educational Technology Centers in Piauí

Nucleus	Municipalities	Number of trainers	Number of schools	Number of linked municipalities
NTE	Floriano	13	50	36
NTE	Greater Teresina	02	51	16
NTE	Oeiras	01	14	10
NTE	Pamaíba	06	47	22
NTE	Picos	08	65	44
NTE	Piripiri	06	41	22
NTE	Regeneration	02	18	15
NTE	S. R. Nonato	02	49	33

NTE	Teresina (Center)	10	95	01
NTE	ValenQa	02	14	10
NTM	Barro Duro	02	08	01
NTM	Luzilándia	02	29	01
NTM	Teresina (Marqués)	15	70	01

Source: Coordination of School and Community Digital Inclusion (CIDEC) - SEDUC - Piaui Formatting by the author

All these investments were made with the aim of improving the quality of education in schools, to enhance the teaching and learning process, as well as to encourage education that takes into account scientific and technological development.[10]

3.2.2 ProInfo: historical data on its implementation in Piauí

The National Program for Information Technology in Education (ProInfo) was implemented in 1997 in the state of Piauí, through a pedagogical proposal of adherence. A commitment to meet the objectives and strategies of the National Program. Through the State Education Department (SEDUC), the state was supposed to set up a commission to draw up a project that would complied with the ProInfo Guidelines (1997). The project had to be specific to educational technologies, in line with the national guidelines of the Ministry of Education (MEC).

In this regard, the state was to define strategies for computerizing public primary and secondary schools. In this process, it was decided that each school unit should present a plan involving digital technologies and education, with a 5-year target. This plan should include didactic and pedagogical objectives, training suggestions for education professionals and a timetable for installation and start-up.

The Teresina Technology Center (NTHE) was created on the initiative of the Municipal Education Department (SEMEC) to implement educational IT in the municipal school system. When the Ministry of Education, through the ProInfo managers, was informed of SEMEC's interest, it sent resources to train teachers and authorized the creation of a municipal center, called the Teresina Educational Technology Center (NTHE), which had to follow the guidelines of the state NTEs.

Created by Decree Law No. 2,794 of June 30, 1999, the NTHE was attached to the Murilo

[10] Available at: <http://www.proinfo.mec.gov.br/site/mapante_sozinho.php

Braga Municipal School. The staff of the center is organized with a coordinator, a secretary and teachers who are specialists in the areas of information technology in education and technologies in education. The NTHE was inaugurated on August 31, 1999, the date on which it began its activities in the municipality, with the aim of training teachers from schools in the municipality of Teresina. With regard to its activities, I would point out that in addition to practical courses involving the use of digital technologies, the center prepares texts, handouts and books for the updating and ongoing training of teachers in educational information technology; it promotes research and suggests initiatives to SEMEC regarding the use of digital technologies in education. These are actions to raise awareness of the need to adopt a new attitude towards the teaching and learning process. They also suggest expanding the capacity to produce knowledge in a shared and emancipatory way. In addition, it aims to provide technical support to teachers, with a view to their day-to-day performance in the school environment. Another important factor included in the initial NTHE project refers to the process of monitoring and evaluating the impact of technology on the educational process and improving the quality of teaching.

According to Castro (2010), the duties of the Technology Center include training teachers in IT courses in education; preparing materials (texts, handouts and books) for the updating and ongoing training of municipal teachers in educational IT; promoting research and proposing initiatives to SEMEC regarding the use of information technology and new technologies in the field of education; preparing distance learning courses to train teachers and other professionals and providing services to the Piauí community when requested. (CASTRO, 2010, p. 14)

Since the beginning of its activities, the NTHE has been under the responsibility of SEMEC and has encouraged changes in the pedagogical practices of teachers in municipal schools. These are changes in the way they teach; encouraging teachers to take part in continuing education courses, pedagogical workshops and pedagogical advice to schools that have a computer lab purchased by ProInfo.

According to Demo's reflections, the training of teachers working with students in grades 6ª to 9ª must undergo significant changes. I believe that in this new educational scenario, teachers need to include activities involving digital technologies, such as navigating cyberspace in order to build and acquire knowledge. This new teacher profile needs to add innovations to the pedagogical project, and this action involves reconfiguring pedagogical practices, based on argumentation and reflection on this practice (DEMO, 2002). Valente's reflections also point to these needs, when he says:

> The issue of teacher training is of fundamental importance in the process of introducing information technology into education, requiring innovative solutions and new approaches to underpin training courses (VALENTE, 1999, p. 19).

In Teresina, the NTHE has the autonomy to set up workshops and courses it deems necessary. According to the coordinator, a major concern is the qualification of teachers in the municipal network, seeking to innovate and meet the needs of the school. One of the NTHE's innovations is the offer of semi-presential courses. The aim of this initiative is to make it easier for teachers to take part in the courses offered by the center.

3.2.3 Broadband in Schools Program (PBLE)[11]

The Ministry of Education (MEC) is responsible for managing the Program in partnership with the National Telecommunications Agency (ANATEL), the Ministry of Communications (MCOM), the Ministry of Planning (MPOG) and the Brazilian State and Municipal Education Departments. It was launched on April 4, 2008 by the Federal Government through Decree No. 6.424, which modifies the General Target Plan for the Universalization of the Public Switched Fixed Telephone Service (PGMU) under Decree No. 4.769.

According to information from the Federal Government, by signing the Additive Term to the Term of Authorization for the exploitation of Fixed Telephony, the authorized operators exchange the obligation to install telephone service stations (PST) in the municipalities for the installation of network infrastructure to support high-speed internet connection in all Brazilian municipalities and connection of all urban public schools with maintenance of services free of charge until the year 2025.

---

11 More information can be found at http://www.mclusaodigital.gov.br/outros-programas

The programme was created with the aim of connecting all urban public schools to the *Internet*, the World Wide Web, using technologies that provide good quality, speed and services that can contribute to universalizing and boosting public education in Brazil. The PBLE is expected to be available until 2025, after its final implementation. The program aims to benefit 37.1 million students in Brazil's 5,565 municipalities.

The laws that govern and guide the Broadband in Schools Program are: Decree No. 2.592, which in its Chapter I, Art. 1 states that for the purposes of this Plan, universalization means the right of access of every person or institution, regardless of their location and socio-economic condition, to the Fixed Switched Telephone Service intended for use by the general public, provided under the public regime. Decree No. 4.769 and Decree No. 6424, which amends and adds provisions to Decree No. 4.769.

3.3 About other federal government programs

New digital technological resources are constantly appearing with the aim of facilitating people's activities, whether in the family, at work, at school or in social relationships. In this section, I will highlight some of the Federal Government and Ministry of Education (MEC) programs created with the aim of digital access and inclusion in education.

One of the most recent is PROUCA[12] , which was instituted by Law No. 12,249 of June 14, 2010. This program was created with the aim of promoting the pedagogical digital inclusion of teachers and students in Brazilian public schools, through the use of portable computers called educational *laptops*. The computer contains a specific operating system and physical characteristics that make it easier to handle and transport, and is designed for the school environment.

During this launch period, five pilot schools located in five cities in Brazilian states were selected as initial experiments. These cities are São Paulo (SP), Porto Alegre (RS), Palmas (TO), Piraí (RJ) and Brasilia (DF).

In 2009, work began on evaluating and consolidating the first five experiences with PROUCA. In 2010, 150,000 *laptops* were purchased and made available to approximately 300 previously selected public schools in Brazilian states and municipalities. Through the program, each school acquired the equipment.

According to Cappelletti (2012, p.4):

> In Brazil, PROUCA is highly anticipated for the benefits it could bring to people's integration into the knowledge society, strengthening skills in critical thinking, problem-solving, creativity, communication, collaboration and autonomy.
>
> PROUCA innovated in its planning strategies, proposing unprecedented work processes in the field of public policies in the area of education. Initially, the Working Group (GTUCA) was created with the aim of advising the MEC, drawing up the proposal for implementing PROUCA. Once the objectives, principles and recommendations were aligned, three pillars were created to support the program: the Evaluation Working Group (Evaluation WG); the Training Working Group (Training WG); and the Research Working Group (Research WG).
>
> Together, the working groups were responsible for proposing different projects, integrating expectations and procedures. This initiative gave PROUCA consistency in its expectations and ideological/theoretical coherence in the intentions and proposals of the different WGs. Both the planned training proposal and the first stage of the evaluation (*basicline*) are being implemented and the work carried out by the research WG resulted in CNPq / Capes / SEED-MEC Notice No. 76/2010. The aim of this notice is to provide financial support for projects characterized as scientific or technological research or innovation clearly related to the use of *laptops* in schools participating in PROUCA (CNPq, 2010).

PROUCA is an innovation in Brazilian education aimed at meeting the social demands of digital access and inclusion, in view of the mobile technology and applications available, in various areas, for audio and video. This initiative was influenced by the One *Laptop* Per Child project

[12] More information can be found at http://www.uca.gov.br/institucional/

conceived by Nicholas Negroponte[13] , who in 2005 presented it at the World Economic Forum meeting in Davos, Switzerland. That same year, Negroponte and his team founded the *One Laptop Per Child* (OLPC) organization.

The program has been much discussed among education professionals, especially teachers, for the possibilities it offers, providing significant changes in didactic-pedagogical planning, in the way of teaching and producing knowledge. On the other hand, as Marques rightly observed, in relation to PROUCA, there is "a need for schools to reformulate their political pedagogical projects and for there to be greater involvement from the entire school community and ongoing teacher training" (MARQUES, 2009, p. 80).

11Researcher, co-founder and director of the MIT Media Lab.

CHAPTER 4

PEDAGOGICAL MEDIATION AND DIGITAL TECHNOLOGIES IN THE TEACHING PROCESS

In order for technology to be used in a meaningful way in the educational context, teachers need to reflect on this reality, rethink their pedagogical practice and build new ways of acting in school. They need to take hold of these technological devices in order to use them as resources for pedagogical mediations that support education and the teaching and learning process, without forgetting, however, the historical and cultural perception of the subjects. In this respect, I take Adams' (2010, p. 37) concept of pedagogical mediations, which states:

> I relate pedagogical mediations to the understanding of education as a pedagogical project that constantly "impregnates" hearts, minds and bodies with motivations, knowledge and critical capacity; and this on an individual and collective level.

Digital technologies used as didactic-pedagogical resources are configured as mediations that enhance the teaching and learning process as they provide interaction between the subjects, represented by the teacher and the student. Based on what Adams mentions, configured as pedagogical mediation, the context of technologies, in the field of education, does not correspond to just being in the environment, more than that, it corresponds to interfering in it, collaborating, "intermediating", causing effective and significant changes in the process of subjects in the learning process. Bringing this understanding to the process of digital inclusion, I understand that the resources of digital technologies promote the emancipation of these subjects, as they acquire a critical vision, develop creativity and autonomy in the search for knowledge and the construction of knowledge.

In addition to the broader concept of pedagogical mediation mentioned above, I would like to highlight the specific role of the teacher. For Masetto (2000), *pedagogical mediation* represents a certain attitude, or posture, of the teacher when they see themselves as mediators of the teaching and learning process, establishing a relationship of exchange. Similarly, Freire (1987) establishes pedagogical mediation in the situation where: "The educator is no longer the one who merely educates, but the one who, while educating, is educated, in dialog with the one being educated who, in being educated, also educates. Both thus become subjects of the process in which they grow together." (FREIRE, 1987, p. 68). I therefore highlight the importance of this process of mediation, when I identify that its main objective consists of fostering critical awareness, the ability to analyze and reflect on the world and the subjects' actions. According to Freire (1987), subjects position themselves as relational beings, able to recognize both their own discourse and the discourse of others. In this relationship, they are able to identify misunderstandings and the particularities of communication between them.

Sharing the same thought, Demo (2001) highlights the great value of the teacher in the teaching and learning process. With regard to digital technologies, pedagogical mediations and the teaching process, it is known that the great difference and importance of educational *software*, or applications created and made available for education, is not in the technological apparatus, in the technical objects, but in the ability of the subjects to use them properly and with defined objectives. "(...) The most essential part of learning is still the teacher - without them we have technology, but not education" (DEMO, 2001, p.1).

In relation to teaching, pedagogical mediation, as necessary, emphasizes the process of interference. In teaching and learning, this can happen without damaging the quality of the relationships of knowledge exchange, information socialization and knowledge construction. In this educational environment, the process of pedagogical mediation and intervention at the right times can guide the process of building knowledge and acquiring the knowledge that really matters.

In this process of teaching and learning, each individual has their own time, their own rhythm. However, it is essential that the teacher pays due attention to the teaching process in order to be able to intervene at the right time for learning. In this sense, the teacher's pedagogical mediation can make significant contributions to the student's learning. In this respect, we understand that pedagogical mediation is a means of guiding and orienting the teaching process so that students can set themselves their own challenges in building knowledge, acquiring knowledge and socializing ideas, enabling conceptions and free choice in the learning process.

4.1 Digital technologies, content approaches and pedagogical mediagoes

The use of technologies in education, from a reflection-oriented perspective to the purposes of digital inclusion, requires the adoption of new pedagogical approaches. These are configured in ways that guarantee the school permanent dialog and attitudes of cooperation with society. In this sense, teachers' didactic-pedagogical practices must meet the objectives of rethinking their initiatives and actions in terms of qualification and the search for knowledge so that they can, in fact, use digital technological resources in the mediation process, giving them autonomy in the democratization of teaching for digital inclusion.

Policies, with an emphasis on education, have been undergoing significant changes in the field of digital technologies. In Brazil, federal public bodies are looking for strategic alternatives that can contribute to educational improvement. In this context, we are faced with the need for awareness, reflection and, above all, effective decision-making on aspects of digital exclusion and the process of including subjects in digital technologies.

This scenario encompasses the individual's socio-economic and cultural development, as well as their relationship with digital technologies and the teaching and learning process. As such, it is of fundamental importance to study the actions of programs arising from public policies of the Federal Government, planned and developed in schools, checking whether or not they enhance the conditions necessary for teaching to be constituted as strategic actions for inclusion and the construction of knowledge with the use of digital technological resources.

From this perspective, digital technologies, especially information and communication technologies, are resources that, if they are not used critically by the subjects, knowing their possibilities and obstacles, can be objects of the ongoing process of social exclusion, creating a scenario that is limited to technical knowledge of digital equipment and resources.

Since the 1990s, digital technologies have been increasingly introduced into state and municipal public schools, given that these technologies have become important resources for the process of economic development and social inclusion in Brazil. In this way, I believe that education must reflect on its pedagogical projects, seeking to adapt them to the demands of society in general, which is experiencing the advances of these digital technologies, without forgetting, however, the humanitarian characteristics that are essential in the formation of subjects. Freire (1996) already mentioned this issue when he stated the need for coherence between pedagogical know-how and know-how. In this respect, Moran (2007, p. 167) believes that:

> The more advanced the technologies, the more education needs human, evolved, competent and ethical people. There is a lot of information, visions and news. Society is becoming increasingly complex and pluralistic and requires people who are open, creative, innovative and reliable.

In view of the resources programmed, the importance of actions to implement digital technology in teaching and learning systems, as well as the seriousness of the objectives for the purposes, it is necessary to establish a process of monitoring and evaluation that shows the results of these actions. How can we prevent the forms of use and access from occurring under equal conditions and from an emancipatory perspective, so as not to increase educational and social inequality and, consequently, exclusion?

According to Pierre Lévy, the issue of exclusion has become important with the advances of cyberculture[14] and points out the exclusionary implications of the growth of cyberspace. Lévy states that the exclusion generated by the dynamics of cyberspace corresponds to the contemporary form of oppression, social injustice and misery. For the author:

> [...] a proactive policy on the part of public authorities, local authorities, citizens' associations and business groups can put cyberspace at the service of the development of disadvantaged regions by exploiting its collective intelligence potential to the full (LÉVY, 1999, p. 235).

[14] A term used by Lévy (1999) to indicate the means and materials of digital communication. Universe of information and human interaction through this cyberspace.

The meaning of digital inclusion is built on the expectation of the constitution of collectivity, in which subjects recognize each other with common purposes and objectives. Castells (1996) points out that the factors of acquiring and processing complex information are linked to the future that is being built. This mutual and continuous influence is contradictory because, on the one hand, it seeks to make life easier for people in the society in which they live and, on the other, it promotes the exclusion of a large part of the population from economic and social life and from the professional market, which absorbs digitally emancipated professionals more easily.

In this sense, education can be configured as a way of contributing to changing the scenario of exclusion of individuals in the world of work. This process takes place to the extent that new practices and new models of social relations, especially those configured as networks, are developed in collective environments, through the introduction of digital technologies configured as auxiliary resources in the constitution of subjects in the teaching and learning process.

It is known, however, that in today's society digital technologies are not yet available to everyone equally, causing tension between those who have access and those who do not. In Brazil, in 2009, data from the National Household Sample Survey (PNAD) revealed that 67.9 million Brazilians claimed to have accessed the *internet* in the previous year. However, even though access to digital technologies is on the rise in Brazil, it is known that this indicator is far from reaching Brazil's population of 201,032,714 inhabitants, estimated according to the most recent data from the Brazilian Institute of Geography and Statistics (IBGE), a fact that interferes in various social sectors and aspects of people's lives (IBGE, 2013).

School units, as institutions, work according to the guidelines of the educational system managed by the Ministry of Education. However, when they act in a focused way, they no longer treat the people they serve as a whole, but rather analyze their needs, taking into account their individuality. In this way, it can be seen that digital inclusion programs, when they are carried out with a focus on the local reality, provide for the subject to be inserted into the job market. The job market, which is currently considered to be very competitive, generally only gives opportunities to those who are more skilled and able to perform activities involving digital technologies.

According to Oliveira (2001), in the process of planning and constructing content in didactic-pedagogical practices mediated by the use of technologies, it is important that there are spaces for debating the relationships between subjects and the network. In this context, I would like to highlight one of the aspects that the author emphasizes:

> In a world of societal inequality and exclusion, where there are few opportunities, not only in the *space of flows,* but also in the *space of places*, for those who are not computer literate, for those who are not up to date with the new information technologies, a logic of exclusion is built around the very agents of exclusion. Thus, in opposition to the net, old and new cultural identities are reaffirmed and built. These resist its discrimination and exclude the *network* themselves, or integrate into it, through activities that are marginal to the values assumed by the wider social system (OLIVEIRA, 2001, p. 105).

The process of digital inclusion and the relationship between individuals and the network thus involves a new culture, a new way of performing and relating to each other. In view of this, it can be seen that public policies, through the process of digital access and inclusion, provide people with the opportunity to be qualified in the use of technological resources, considering that, as a result, professionals are able to face up to the competition in the world of work. Even though governments fund and implement programs and courses that enable digital access and inclusion, it is assumed that most of the poor population does not even have access to information about these courses.

According to Freire (2005), only man is capable of transforming his current state, based on a critical view of what he is suffering. Therefore, it is necessary to conceptualize your reality and decide on concrete objectives for your path, not only in relation to the professional sphere, but above all in relation to life itself. Thus, as Freire states, "[...] in the relationships that man establishes with the world there is, for this very reason, a plurality in his own singularity" (FREIRE, 2005, p.48).

Based on what Freire suggests, we can also see the thinking of Lévy (1999), when he points out that man's relationship with the world is only imaginable if the aim is to use the digital environment to develop regions where the socially disadvantaged find themselves, giving deserved value through projects, as well as exchanges of experiences and knowledge. This requires the participation of the school community, as well as more forceful action by public bodies with public policies.

With regard to digital exclusion, whether it's the lack of access to technological resources or the promotion of digital media skills in order to compete in the job market, I understand that this issue can be seen in different ways. Digital inclusion means promoting access to productivity and, above all, to new models of social relations. However, for the process of including people, I understand that it is not enough to prepare them in the field of digital technologies or to adapt strategies for obtaining technological resources, because only part of society is in a position to acquire these technological devices.

There is no doubt about the importance of using digital technologies in education. However, these technologies are a major challenge because we know how they are often introduced into school environments. Therefore, it is not enough to defend the insertion and use of digital resources if they are not adapted to the school context and by trained professionals who are able to prepare everyone involved to live in a society that is now conceptualized as digitally technological.

According to Costa (2006), the process of inclusion involves economic indicators, which means the need for financial conditions for access to digital technologies; it also involves cognitive indicators, which means having a critical vision and autonomy to use and appropriate the new digital media; and technical knowledge indicators, involving operational skills for programs and access to the *Internet* system.

From this, I understand that inclusion is a broad process that should not be just a simple action of technical training of applications and skill in the use of technological resources, but mainly involves awareness of knowing how to use them in appropriate contexts, involving decisions on how to use them and what to use them for.

Faced with this scenario, it is imperative to make good use of the improvements proposed by the advances that have been made, especially by the process of globalization, in order to prepare people for the humanitarian and professional fields, which are increasingly demanding of people, in a process that is based on school.

4.2 Pedagogical practices: new technical objects require new skills

Over time, the technical objects[15] used as teaching resources by teachers in their teaching practices have changed and adapted to new configurations and, I might even say, to the demands of the teaching and learning process. With this in mind, I intend to take a historical look back, not at all, but at the main teaching resources, configured as technical objects, used by teachers since the Gutenberg press. I'm focusing on the mid-20th century up to the present day, 2013, because I consider this to be a period that has seen a significant change in the teaching and learning process in everyday classroom life.

Here, I point out the concept and evolution of the use of technical objects in order to gain a better understanding of what a teacher will value when planning and choosing teaching resources to work with educational content in the teaching process. In this respect, I would highlight what Santos (1994) says about technical objects: "Pre-existing objects are aged by the appearance of more technically advanced objects, endowed with superior operational quality" (SANTOS, 1994, p. 50). For the author, the record of the evolution of time teaches us about the meanings and conceptions of objects and things in general created and used by individuals throughout their lives.

Santos also points out that it is always necessary to recreate technical objects in order to adapt them to new realities, considering that time passes and, in the process, objects and things change and/or transform in an evolutionary dynamic.

In 1981, the American company IBM launched its first personal computer (PC) and, in the same decade, the first PCs appeared in Brazil. The evolution of these technical objects gained momentum in the 1990s with the refinement of their technology. However, public access only began

[15] Technical objects aredefined here as artifacts created to meet the needs of human beings and the environment in which they live.

in 1994. The 1990s also saw the creation of advanced *software* and the *Internet* or *Word Wide Web* system, considered by many to be a powerful communication resource from the start of the 21st century, developing at a faster rate than any previous technological communication resource.

According to Paiva (s/d, p.8), access to the World Wide Web in Brazil began in 1991 with the creation of the National Research Network (RNP) by the National Council for Scientific and Technological Development (CNPq). From then on, teachers began communicating via the network and many other activities that contributed to teaching and learning.

The *Internet* is currently an important technological resource for the educational system, taking on the role of an environment for researching and socializing content, ideas and information, much used by the new generation. The ability to research, to form study groups and discussion forums, to produce and socialize content in an *online* and/or virtual format are potentially extraordinary resources in the hands of Internet users. I therefore believe that, when it comes to education, teachers need to be aware of the profile of this new student, who is familiar with and even masters these new digital technological resources. According to Freitas and Leite (2011):

> Teachers who have dealt with and will continue to interact with these new generations arriving at schools need to learn how to get to know them and challenge them pedagogically; and for this to happen, they need to receive adequate training to accompany the continuous development of technology (FREITAS and LEITE, 2011, p. 33):

One thing seems certain: society is currently witnessing an avalanche of new technical objects and digital technological resources for use by education professionals, which will include more advanced means every day so that they can mediate the teaching and learning process in a creative and more effective way. In this scenario, I realize that it's not just about the technique or the technology, but about what these resources can do for the educational system and, above all, for the subjects involved in the process.

In this respect, I would emphasize that these resources, if properly applied, with an adequate structure in the schools, technical staff available for support and also technological training for teachers, will all contribute to the effective success of the educational system. Lopes (2010), when referring to art and technique, puts it in terms of:

> beyond its utilitarian perspective and as a possibility of producing innovative learning and pedagogical mediation, as well as [...] the potential of technical activity in the educational context as a motivating and empowering element of learning and creativity (LOPES, 2010, p. 16),

Intellectually understanding this new reality puts people in a challenging situation: understanding and living with facts that require competence to produce epistemological information, as well as adequate knowledge based on experience. At this point, Santos proposes that we think about the understanding and construction of things:

> We know that the permanent is not permanent because the successive visions made possible by knowledge dismantle our construction of things, even those we considered eternal. And we also know that we can't encompass all of today, but it is our task, nevertheless, to seek to understand it. (SANTOS, 1994, p. 44)

As a reflection on what the author is referring to, we could say that some important technical objects, adopted in the old days, are still used today by teachers and students in the development of everyday school activities. Examples include the 15th century printed book, the good old notebook, the pencil, the pen, the paintbrush, the blackboard, among others. Although nowadays new technical objects, some technically advanced, are at the disposal of education, in some places objects from the old days are still extremely important in action systems, because they fit in with their current realities. In this sense, I recognize that technical objects are considered systems that integrate and interact with action systems. From the point of view of Santos (1994):

> On the one hand, the systems of objects condition the way in which actions take place and, on the other, the system of actions leads to the creation of new objects or is carried out on pre-existing objects. This is how space finds its dynamics and transforms itself (SANTOS, 1994, p. 55).

Therefore, I understand that changes in these realities permeate not only the conception of the teachers and/or managers of the school system, but mainly the understanding and actions of the Public System, at a higher level, as the holder of Power.

One of the advantages of thinking about education in a creative and innovative way, with these new technical objects and digital technological resources of today, is that they require a new model of acting and developing activities in the educational system, with the main and imperative focus on this new generation of students, with different characteristics and profiles, from the point of view of their needs and priorities arising from contemporary times.

According to Couto (2013), we can currently define a model of education with different characteristics, known as Education 3.0[16] , because:

> The mere presence of technical objects in the classroom does not necessarily mean innovation. It could even be a big step backwards. The computer alone does nothing. In order to use digital technologies in an innovative way in teaching practices, we need to solve three problems simultaneously: improving the technological infrastructure; improving access to the network and adequately training teachers in digital culture. These three points actually emphasize that when we talk about digital technologies we are no longer talking about machines, but about connected people, doing incredible things because they are together, working in partnerships, collectively. If people aren't connected and don't have the freedom to discuss and create, nothing will change in education (COUTO, 2013, n.d.).

Thus, in the educational system, the process of teaching and learning has established new skills over the years, as well as the realization that, especially today, we are living in a period in which knowledge and skills about technology and technology are configured in a methodology of mutual influence. In the system of ideas of the educational process, I emphasize the understanding of "why do" and "why use", even though I understand that, in the process of using these technical resources, objects and things, one cannot exclude what concerns "how to do" and "how to use". According to Lopes (2010):

> In general, technique is understood from a utilitarian, pragmatic point of view, directly applicable to a reality or context. In fact, faced with so many technological resources and instruments produced by human beings throughout history, it is understandable that so much emphasis is placed on products and less on the symbolic processes involved in their development and use. From this perspective, knowledge becomes a thing and acquires market value (LOPES, 2010, p. 18).

For the author, in general, the acquisition of advanced technological objects makes the subject feel valued, giving them a position of prominence. The understanding of technology and the handling of objects has influenced the way "we live, dwell, produce, consume, work and communicate" (LOPES, 2010, p. 19).

In this respect, there are widespread issues at the heart of current discussions about the use of technical objects and digital technologies in teaching practices in the daily life of educational institutions, a fact that differs in conduct and conceptions, which is why I was inspired to produce this work.

4.2.1 Digital technologies in the teaching process

Current theories on the use of digital technologies in the teaching process lead me to take a critical look at the relative meaning of reality and its relationship with the theories and models of discourse that have been used in this regard. Today's reality is distinguished by its complexity and diversity in schools nationwide. In this sense, this diversity of aspects allows for a variety of interpretations. In this context, I emphasize the reflections of Santos (1994), when he says:

> The history of man on Earth is the history of a progressive rupture between man and his environment. This process accelerates when, at almost the same time, man discovers himself as an individual and begins to mechanize the planet, arming himself with new tools to try to dominate it. Artificialized nature marks a major change in the human history of nature. Today, with technoscience, we have reached

[16] According to Couto (2013), Education 3.0 is people technology, which integrates people through digital technology.

the supreme stage of this evolution. (SANTOS, 1994, p. 5)

You need to be in a constant learning process, because with every new technical object and new technology that emerges, there is a need to get to know them, their purpose, their proper use and their handling. This is a fact that stands out in our time, as things and objects change and evolve rapidly. These are innovations that, according to Santos, always place us at the stage of ignorance, but also in front of the opportunity to renew our knowledge, that is, to learn everything anew, which is an important option in the process: "Never, as nowadays, has there been a need for more and more competent knowledge, thanks to the ignorance induced by the objects that surround us, and the actions that we cannot escape". (SANTOS, 1994, p.45)

In other words, the difficulty of always being up-to-date, computerized and connected forces human beings to seek out new knowledge, to interact, to be part of a new reality that is presented in the world of technological innovations. Otherwise, human beings accept the condition of being renegades and outside the evolutionary context of potentially advanced things, concepts and environments. Often, this hard choice depends not only on the entrepreneurial and creative spirit of the individual, but mainly on the conditions in which they find themselves.

Furthermore, with regard to education and the new models of systems of technical objects and systems of actions, Perrenoud, quoted by Freitas and Leite (2011, p. 32), points out: "[...] the evolution of the school transforms the profession of teacher decade after decade, by a double movement: growing ambitions and increasingly difficult conditions of exercise". I can complete the thought with the statement that "[...] a basic technological culture is also necessary in order to think about the relationships between the evolution of tools (information technology and hypermedia), intellectual skills and the relationship with the knowledge that the school wants to form". *(Ibid)*

In this technological scenario, hypertext emerges as a source of research and content production, which according to Levy (2004) is a set of nodes linked by connections. The nodes are content presented in words, images, graphics, sound sequences and more complex texts which, in turn, can also be configured into new hypertexts. I would stress here the importance of teachers using this source to produce content and socialize information in their teaching practices.

4.2.2 Historical retrospective of technical objects used as teaching resources in the classroom

Here I propose to make a historical survey of the evolution of technical objects and the pedagogical use of these objects by teachers, considering everyday school life as a space configured from an inseparable set of systems of objects and systems of actions.

To support this proposal, I bring to the educational system what Santos (1994) says: "Systems of actions also do not occur without systems of objects. Space today is a system of increasingly artificial objects, populated by systems of actions equally imbued with artificiality" (SANTOS, 1994, p. 44). In this respect, I believe that the difference in the use of these objects in the school space is established in the manner appropriate to the place and the subjects of that place, their conceptions, their meanings and the meanings given to them.

If, in the past, the technical objects that defined teachers' pedagogical practices and their way of working with the content taught in the classroom were important for that context, today, as these objects change and become more popular, they are used specifically to meet the needs of regions, localities and their unique and plural characteristics.

Recognizing the advances in computer technology and the *Internet,* which, in their current formats, are full of advanced features, expanding the possibilities of use and connection on a large scale and worldwide, it is necessary to remember other historical moments in which techniques emerged that are still current and used. This is the case of the written press.

The year 1442 is a period marked by a major technological revolution in human history. In this regard, I would highlight a great figure in the history of the 15th century, the inventor Gutenberg, who created an extraordinary machine: the printing press, with the Bible being the first printed book.

Then came the printed book on a larger scale and the possibility of socializing content and information. Before the Gutenberg press, books and intellectual productions were produced by extremely skilled professionals, called copyists, who drew letters and formed words, sentences and texts to make reading objects in the format of the time. Today, the book faces other challenges since the introduction of more advanced objects such as the computer into our society. However, not everyone had access to this object and, in some places, access was forbidden by the state and the church. Today, books are also becoming digital, allowing greater democratization of access, but printed books are still valued, at least until now.

4.2.3 Audio and video technologies

As a specific point of view, I emphasize the historical events that marked epochs and changed the way of being and being in the world. Here, I will focus on sound and video reproduction, a major technological innovation that emerged at the end of the 19th century. The first technical objects created were limited only to sound reproduction, but later came image projection technology and, subsequently, technical objects that simultaneously reproduced sound and projected images. With the technological innovation of sound recording and reproduction, it was increasingly possible to bring recorded content and information into the classroom, facilitating the teaching process in the pedagogical practices of the more daring teachers who appropriated this technology.

However, for many others, technology and innovative technical objects were technically sophisticated and difficult to access. In this sense, as technical objects evolved, for some of the teachers this was seen as a harbinger of loss of job and social position.

Of all the audio and video technologies, including cinema, radio and television, these were the most socializing, but their impact on formal school education, with the exception of television, was not as great as expected. But it is worth noting that important programs developed in Brazil in the 1950s and 1960s, for example the Basic Education Movement (MEB) and the Popular Culture Movement (MCP) used radio to popularize and massify the educational process, an instrument that is still used today in several Latin American countries, along with television.

Television in the regular school classroom takes on a new dimension when it is used to view recorded videos, which have become part of the teaching materials of major publishing houses. With the emergence of digital culture at the end of the 20th century, the broadcasting of audiovisuals migrated to *CD-Roms* and *DVDs*. Currently, the *Pen Drive*, *data show, notebook* and others stand out as technical objects for this purpose.

With the emergence of new technologies, the education system and, more specifically, schools are trying to give teachers access to these resources, as well as adapting the new technical objects to pedagogical practices in the teaching process. This constitutes experience and innovation in the process of improving teaching and the mediation between teacher and student, starting with the singular objective of transmitting content. Therefore, sound and image technologies were added to the printed book in everyday classroom life. In this respect, I can highlight the importance of this technology from the great revolution in teaching that began in 1878, the 19th century, with Thomas Edson's invention of the phonograph[17] , an invention that made him known worldwide. Then came the gramophone with its recording on disks and then the magnetic tape (PAIVA, s/d, p. 5).

Computer and communication system resources have evolved in surprising ways. In this sense, the post and telegraph, the computer and its peripherals make up and integrate all the technologies of writing, audio and video, mediating communication and the relationships of individuals in society and, above all, their way of being in the world.

I'll mention here the technical objects that have stood out over time: the Gutenberg press, the typewriter, the audio and video recorder, the radio, television, the *slide* projector, the video projector, the multimedia *Data Show,* among others.

Since the 21st century, the *Internet* system - specifically through Web 2.0, and now the semantic web (ADAMS et al, 2013) - has given users new possibilities. From being a consumer of content, users are transformed into producers and learners of technology, allowing them to experience the use of communication in diverse and enriching experiences. From these technological resources come the means of relationships and search resources configured in networks, such as *Google* (1998),

---

[17]**Phonograph: An** object with technology that worked on the basis of sound and made the recording diaphragm vibrate. While a cylinder covered in tin foil rotated over the diaphragm needle, the needle made cuts in the tin foil, which varied according to the sound. When the recording was complete, the recording needle was replaced by another, which, rotating again in the cylinder, reproduced what had previously been recorded.

Orkut (2002), *Blogs* (2002), *Skype* (2003), *Facebook* (2004), *Podcasts* (2004), *Youtube* (2005), *Twitter* (2006), *Instagran* (2010), systems that have revolutionized the way human beings socialize and relate to each other. Available to Internet users is *Wikipedia* (2001), the first virtual encyclopedia with worldwide access, built collectively by *Internet* users on a global scale. It is a model that competes with the luxurious and world-famous printed encyclopedias, created in the 18th century, which are the objects of research and reading. As far as *Wikipedia is concerned,* I would highlight the fact that users can also be authors of content, "reconstructors of knowledge" (Demo, 2001), as well as disseminating their intellectual productions, interacting with texts, audio and/or video.

4.2.4 Digital technologies : new challenges, new skills

The computer has become established and expanded in society due to some of its attributes that have evolved, such as: the reduction in size and cost; the wide variety of *software over* time to cater for the various areas of knowledge; the increasingly attractive *design* and the change in users' attitudes towards the new technology. The opportunity to access and purchase computers and, above all, their use in education goes through several stages, starting with access and use for teachers and students in well-regarded private sector institutions. Later, this access was extended to public schools, making this important technical object a mass object and universally used.

Nowadays, in terms of the teaching and learning process, teachers are taking ownership of this technically advanced object, with the computer becoming an increasingly integral part of teaching practices in everyday school life, given the new profile of students. In this context, new skills, competences and innovative pedagogical concepts are being demanded. For Couto, this means

> Education 3.0 [...] brings digital technologies into the classroom to stimulate the production and exchange of knowledge. The emphasis should not be on technical objects, their environments and applications, but on interactions, exchanges and collective doing. So the classroom becomes any environment where people connect with each other and create, find solutions to their problems, face their dilemmas collectively. Where there are connected people, there is teaching and learning mediated by digital technologies. The teacher is no longer the one who transmits a certain amount of ready-made knowledge. Being a teacher in a digital culture means coordinating, guiding and encouraging collaborative and increasingly personalized learning. It is no longer a question of the same task for everyone at a given time. The teacher is now the one who coordinates activities around a problem, or certain problems. In this way, many groups, at different times, can work together. Each teacher, each student, can open a research front and everyone can share doubts and discoveries. The continuous exchange of experiences becomes a fundamental value of Education 3.0.

I emphasize that these paradigmatic shifts are a fundamental part of the new models of life in society that are emerging and from which teaching and learning cannot be left out. These are necessary transformations in the educational system which, once assimilated, tend to go beyond the school and the classroom and extend beyond these environments to the communities, significantly interfering in the process of social transformation.

In Brazil, in terms of education, there is still a tension between the process of digital inclusion and exclusion, which is often configured by adherence to these new technological resources and, conversely, rejection of their use and adaptation; or even by the impossibility of accessing them.

In the evolution of pedagogical practices and the use of technical objects seen as important tools in mediating the teaching and learning process, I would like to highlight the computer and its peripherals, which were used by teachers in higher education schools in the 1980s. I recognize that, even so, a huge portion of the population in Brazil today does not take advantage of the various possibilities of information technology.

As far as the choice of resource for educational use is concerned, it should be understood that this choice depends on the context, the way the school is organized, the learning objectives, the resources available on site, the infrastructure, the pedagogical preferences and methodology used by the teachers, it also depends on the students, their life experience and the knowledge they bring with them and, most importantly, the availability of technology in the form of *hardware* and *software*. As part of the process, the way in which these resources are used in the organization of the classroom space is equally important, as is the way in which groups are organized and integrated, and interdisciplinarity. So it's up to teachers to rethink their way of conceiving the world and things in order to adapt to new models and appropriate methodologies, and to use them in everyday life as tools rich in possibilities.

I emphasize the fact that the history of technology and technical objects in education, more specifically in pedagogical practices, could not be linear in Brazil, where social differences prevent these technological resources from being available to everyone. Many obsolete technologies, such as *slide* projectors and VHS video cassettes, for example, never reached certain schools in the country.

In the education system, but precisely when we refer to teachers' pedagogical practices, probably not everyone will have access to computers and the *internet*. However, we need to be aware that neither the printed book, nor the computer, nor the most advanced *software* and/or technical objects used in this context will solve all of education's problems in the teaching and learning process. I would emphasize that the difference that is made in this educational scenario depends on the intentions behind the use of technology and technical objects, the conceptions about them, and the vision and objectives of the individuals who are in charge of education.

CHAPTER 5

METHODOLOGY

In order to support the definition and support of the methodology of this research, I looked for a basis in the works of the authors Minayo (2011), Malheiros (2011) and Gil (2010), whose theories describe the subject in a clear and didactic way, which helped to improve my understanding of the subject. In this chapter, information is given on the methodological procedures for collecting the data considered important for the analysis, treatment of the results and conclusions about the research problem, as well as the previously defined objectives.

In this sense, I was interested in investigating the conditions in which digital technologies are being used by teachers, the influences on didactic-pedagogical practices and conceptions of these resources. As participants in the research, I chose to interview four teachers who have been using digital technologies and the ProInfo laboratory installed at the school since 2002 and one teacher who does not use digital technologies in his daily activities, but has attended the computer lab when it was monitored. Also taking part in the research was the coordinator of the Teresina Technology Center, which is responsible for running the program in the municipality, managing training actions for teachers in municipal schools, providing technical assistance and monitoring schools that have signed up to ProInfo. The sample totals 6 (six) respondents, who took part in the interviews carried out in the empirical field, which ensured a broader view of the reality of the school under study.

In this respect, during the research process I worked from the perspective of understanding and reflecting on didactic-pedagogical practice mediated by digital technologies, highlighting the way in which the research subjects conceive, give meaning to and narrate their experiences with DM and the computer lab (ProInfo). In the interview script, organized with semi-structured questions, three thematic axes were pre-established in order to define the dimensions of the research. The **first theme** involves teachers' conceptions of and experiences with digital technologies; the **second theme** concerns training initiatives, including competencies and skills for using these technologies; the **third theme** is specific to the computer lab acquired by ProInfo, the relationship between teachers and their pedagogical practices developed in this lab.

5.1 Qualitative aspects of the research

The work is configured within the scope of qualitative research, as it seeks to obtain from the subjects, in the field of subjectivity, their conceptions and meanings given to the object under study. I emphasize the role of teachers in their pedagogical practices in the teaching process mediated by digital technologies in the construction of knowledge and the acquisition of knowledge. I try to highlight the activities that guide and enhance the path towards digital emancipation and the appreciation of the social reality that is established in this school space.

According to Oliveira (2002), quoted by Marques (2009), the qualitative aspect of research is justified:

> The fact that the qualitative treatment of a problem can be chosen by the researcher is an appropriate way to understand the cause and effect relationship of the phenomenon and consequently arrive at its truth and reason. Furthermore, research using a qualitative approach can easily describe the complexity of a given hypothesis or problem, analyze the interaction of certain variables, understand and classify dynamic processes for social groups, contribute to the process of change, creation or formation of opinions in a given group and allow for a greater degree of depth in interpreting the particularities of individuals' behavior or attitudes (OLIVEIRA, 2002).

At this juncture, it is important to emphasize that the qualitative approach is based on the construction, vision and interpretation of the reality of the subjects who interact with each other at the level of the cultural, social and educational aspects of the environment to which they belong. Another aspect of qualitative research includes the specific realities, their implications and the changes that take place in this space.

I would like to point out that, because the research was carried out in this way, it involved a lot of information collected during the investigation process, ranging from official documents, reading and access to computer databases, *websites,* books, academic papers, among others, a fact that required the researcher to be more involved with the proposed work.

5.2 Research clipping

In this topic, I present the focus of the research, configured as a cross-section, the design of the field, the subjects participating in the research, the method of analysis, the procedures and the strategies for collecting data and information, as well as the activities developed from visits to the school under study for observations and notes in the field diary, activities that preceded the interviews.

My main interest is focused on the pedagogical practices of teachers at a municipal public school in Piauí who use digital technologies as mediating resources in the teaching process. In this sense, I am focusing on computers and the *internet,* which have become important digital resources since the implementation of the National Program for Information Technology in Education (ProInfo), which provides computer labs and digital resources for teachers and students to use.

As far as the period defined for the study is concerned, I opted for the start of the ProInfo laboratory installations at the school, in 2002. In 2008, the Federal Government launched the Broadband in Schools Program, with the aim of connecting all urban public schools to the *Internet,* using technologies that include quality, speed and services for the innovation and development of education in the country. The MEC and the National Telecommunications Agency (ANATEL) are responsible for managing the Program, in partnership with CONSED, the Ministry of Communications and the Ministry of Planning and the Brazilian State and Municipal Education Secretariats.

The research and the data collected in the field were taken from a public school located in the municipality of Teresina, in the state of Piauí. This data is the result of visits, informal dialogues with teachers and pedagogues, observations recorded in a field diary and interviews with 06 participants.

The school is linked to the Municipal Department of Education (SEMEC), which offers the community of Teresina the highest level of primary education, from 6th to 9th grade. It has 30 teachers and 892 students in the morning and afternoon shifts. In the evening, it caters for around 120 students from PROJOVEM - a federal government program - which aims to promote the social inclusion of young Brazilians aged 18 to 29 who, despite being literate, have not completed elementary school, seeking to reintegrate them into school and the world of work, in order to provide them with opportunities for human development and the effective exercise of citizenship.

In addition to curricular activities, the school offers courses in IT, Contemporary Dance, Ballet, Guitar and Choir. It also runs social projects called School Support, Environmental Education and the School Violence Project. It also carries out various activities involving the surrounding community, with the aim of promoting school-community interaction. Currently, the ProInfo laboratory is being used to support "Mais Educaçao", a federal government program. At the school, the course taught is computer science and there is a person from the community who manages the time the students spend in the lab.

The school's current physical structure consists of 25 classrooms, a teachers' room, a principal's office, a secretary's office, a pedagogical coordinator's office, a video room, a library, a music room, a dance room, a computer lab - ProInfo, a science lab, a cafeteria, toilets, a storage room and spacious open areas for socializing and leisure activities.

The participants in the research are teachers from the above-mentioned school. In this process, priority was given to those who use digital technological resources to plan and carry out their didactic-pedagogical activities. The teachers were chosen through an initial questionnaire applied at the school. From then on, I obtained information about the use of digital technologies by the teachers, and there were also informal dialogues with some teachers who were willing to cooperate with the research. This strategy served to reinforce the field diary and provide a basis for the research results.

I obtained information from the school's managers about the initiatives and management strategies regarding the process of joining ProInfo. The managers are the Director General, the ProInfo coordinator at the school and the pedagogical coordinator. In this respect, I would like to make it clear that this item is intended as a form of reinforcement to support the research.

The manager of the Teresina Technology Center (NTHE), which is responsible for digital inclusion policies and actions in the municipality, also took part in the research, with the aim of obtaining data and information on the management of ProInfo.

5.3 Analysis method

In this work, which is configured as a case study, I used content analysis as a method, based on the teachers' conceptions of digital resources; the initiatives and investments to qualify teachers, in view of the changes in the educational scenario after the adoption and implementation of ProInfo; the suitability of educational environments; the construction of knowledge; the process of qualifying subjects for digital technologies in education, with an in-depth look at the didactic-pedagogical use of digital resources for the teaching and learning process.

Regarding the qualitative approach and content analysis as a method, these are fundamentals that fit in well with my research proposal, because it involves the concepts, motivations, initiatives and attitudes of the subjects being researched. According to Minayo, qualitative research works with:

> [...] a universe of meanings, motives, aspirations, beliefs, values and attitudes. This set of human phenomena is understood here as part of social reality, because human beings distinguish themselves not only by acting, but by thinking about what they do and interpreting their actions within and from the reality they experience and share with their fellow human beings. (MINAYO, 2011, p. 21)

In this sense, I seek to be guided by the qualitative aspects of the investigation, to understand the relationships between the subjects of the research and the technological resources made available for the development of their educational activities in the construction and socialization of knowledge for the formation of the subjects.

According to the author, the meaning of "understand" is in line with the objectives of qualitative research, which aims to understand and interpret human phenomena in their social relations (MINAYO, 2011, p. 24).

5.4 Field procedures for data collection

In this research, I propose a methodology aimed at answering the problem and the proposed objectives. As instruments for data collection, I opted for bibliographical and documentary research, with readings in official documents that correspond to legal measures of public policies of the Federal Government for Brazilian public schools, more specifically, documents referring to digital inclusion programs.

In the empirical field, I decided *apriori to* use, as data collection procedures, instruments that could effectively contribute to the construction of theoretical assumptions that demonstrate and justify the educational reality of a municipal school in Piauí with regard to teachers' pedagogical practices and digital technologies, after the inclusion of ProInfo.

According to Gil, for procedures in the empirical field it is possible to use strategies such as interviews, questionnaires and observations (GIL, 2010 p. 57). In this respect, I tried to combine techniques ranging from direct observations, focused informal dialogues and the application of questionnaires containing questions in the context of the problematization, to help select the sample. After this phase of selecting the subjects, semi-structured interviews were carried out using a script previously organized into thematic axes, to make it easier to identify the dimensions and units of meaning. A field diary was used to record relevant information about the object of investigation. In this sense, I would emphasize that studying the empirical field allows us to observe important details, understand the relationships that are established between the subjects, as well as contributing to a deeper understanding of the issues considered to be the focus of the research.

5.4.1 Describing the Field Diary entries

Research with a qualitative approach requires the researcher to take a detailed look at the object of study in order to obtain, as completely as possible, the real situation and the conceptions about the experiences of the subjects being investigated. In this detailed search for information about reality, it is extremely important to observe the details. In this investigation, I have tried to describe the data collected in detail so that nothing goes unnoticed as far as possible.

Therefore, the aim of this section is to transcribe the path followed during the research process in the empirical field and record information obtained through the strategy of observation, focused informal dialogues and transcribing information pertinent to the object of study in the Field Diary.

I begin my work by contacting the principal of the municipal school selected to be the empirical field of the research. I arranged a visit to start a relationship of cooperation and support for the proposed work. Once this was done, I went to the school and met the principal, to whom I explained my research work, emphasizing the points that I believe to be fundamental: the guarantee of confidentiality, the seriousness of scientific work, the object of study and the research objectives. Immediately after my presentation, the principal referred me to the school's pedagogue, who would provide me with the necessary support during the investigation.

On my first visit, I noticed that the establishment was in line with the standards of a municipal public school in the state of Piauí, with similar characteristics to all the others. Even though it is in a privileged location, close to the center of the capital and on one of the main avenues, the physical structure is clearly in need of a good renovation. However, what struck me was the atmosphere of good organization that I could observe. For example, the small number of students outside the classroom, the silence in the corridors, in the common areas, the security guards positioned at the main entrances and the staff carrying out their daily activities in a normal atmosphere.

While I was waiting for the teacher, a distant period in my professional life came to mind, when I started teaching in public schools in the south of Teresina. At that time, the teaching resources were the teacher's knowledge, his skills and abilities, his commitment to education, his creativity, the textbook, the blackboard and the chalk. In the school at the time, all that was available in terms of technological resources was a stereo with a tape recorder attached, which was widely used by teachers of foreign languages, physical education and art education.

I tried to develop a good dialog with the pedagogue, because I knew that the success or otherwise of my field research would depend on this contact. The teacher was receptive and willing to help me. In the first conversation, I was told that the computer lab I had obtained through ProInfo was also currently being used by the federal government's "Mais Educaçâo" program[18] . Something that also caught my attention was the information that not all teachers use the computer lab to teach their classes, because with the acquisition of mobile technology equipment, such as multimedia *datashow* and *notebooks*, made available to teachers, they currently prefer the classroom.

I was informed that the students attend the laboratory through the "Mais Educaçâo" Program with the support of a person from the community, who only has the technical knowledge to handle the equipment. This person is selected and sent to the school by the Municipal Education Department (SEMEC). Activities in the laboratory are extra-curricular and take place three times a week, with afternoon students attending in the morning and evening students in the afternoon. In their testimonies, the teachers highlighted the students' enthusiasm for the fact that the lab has an *internet* system. However, they were apprehensive because they felt that this person lacked pedagogical knowledge and an activity plan.

I met the Physical Education teacher, who was also interested and willing to cooperate with the research. She even revealed that she uses her cell phone in some of her activities and that in 2002 she had taken part in the training promoted by the Teresina Technology Center (NTHE) when the ProInfo Laboratory was being set up at the school.

After my first visit to the school, I went back three more times to get a better look at the school environment and to administer the first questionnaire (Appendix A) to the teachers. I would like to point out that not all of them were willing to fill it out. However, on the other visits, I spoke to some of the teachers personally and obtained their commitment to take part in the interviews if I so wished. I have to confess that I was relieved, as this meant that I would have the number planned and defined earlier in the research project.

At the school, the computer lab consists of 30 computers and a *Weriless* system for network connection. However, only 10 to 15 computers are working, still in a precarious way, a fact that discourages the teaching community from using it, since one of the complaints is that they don't have

---

[18]The Mais Educaçâo Program, created by Interministerial Ordinance No. 17/2007, increases the educational offer in public schools through optional activities, such as computer activities.
For more information on the program, visit: http://portal.mec.gov.br/

a monitor, technical support or a computer maintenance and replacement program. Some teachers said that the *internet* system is slow and sometimes the network crashes when all the equipment is connected, generating feelings of demotivation and annoyance among users.

One fact that is very evident is the absence of someone to stay in the lab so that, when the teacher arrives, everything is ready for the start of the lesson. The teachers claim that they spend a lot of time turning on the equipment, checking that everything is working properly, etc. "*After all that, a good part of the lesson has already gone*" (P2), says one teacher. Regarding this fact, some teachers reveal that in the beginning, after the computer lab was set up, the NTHE had a monitor who always helped them and the equipment was inspected frequently by a technician.

However, over the years, these professionals have been laid off. This has contributed to teachers leaving the lab, as a number of factors have prevented and discouraged teachers from staying in the lab, such as equipment with technical problems, outdated configuration and a small number of machines in operation.

I then began interviews with the coordinator of the NTHE and with five (5) teachers from the school, selected after discussions with some of them and based on the results of the first questionnaire applied at the beginning of the research, a document which is included in the dissertation as "Appendix A". To take part in the research, I chose 4 (four) teachers who, in some way, use or have used digital technologies in their classes and/or in the computer lab, and 1 (one) teacher who does not use digital technologies, but has been to the lab since it was first set up. This brings me to 6 (six) respondents, the number proposed in the research project.

My first interview was full of surprises, very positive and enriching. The interviewee was enthusiastic about talking about technology and recounted her teaching experience using digital technologies. I was then able to see that I had been mistaken in my first assumption about the use of technology in teaching practices at the school. After long conversations with some of the teachers, I realized that one of ProInfo's biggest problems is the absence of a monitor, the conditions in which the laboratory is used and the fact that the equipment is fixed, considering the mobile technology available today: multimedia *datashow, cell* phones, *smartphone*, *ifone, ipad* and their various applications.

Another fact refers to the fact that, even though more than ten years have passed since the program was set up at the school, there are still teachers who don't use the laboratory because they don't feel able to handle the equipment. In this respect, they demonstrated the importance of the monitor to support the didactic activities planned to be taught in the laboratory. Others said they didn't feel very comfortable because they believe that most students know a lot about digital technologies. They point out that in the beginning the lab was more frequented by teachers, because at that time the equipment was new and there was a monitor provided by the Technology Center of the Municipality of Teresina (NTHE) to give technical support in conducting activities, especially those involving *internet* access. In this context, teachers are concerned that students sometimes divert their attention from the planned activity and access other content on the *web*. This often causes stress, as the teacher has to constantly monitor and control the students so that they don't deviate from the proposed activity. In this case, the teachers make it clear that the monitor is very much needed.

Because today's students have cell phones of different models, *smart phones* with applications for accessing the *internet,* some teachers prepare certain didactic activities including the use of these digital resources. In this respect, I was able to see some very positive experiences reported by some of the interviewees.

I sensed a certain expectation on the part of the teachers regarding the announced purchase of a *tablet*. This was "*a promise from the mayor of the capital during the last election campaign",* says one teacher. I could feel the anxiety of some of the school's teachers waiting for this technical object. They showed that they were confident, but said they didn't know when the equipment would be available. In the view of some teachers, the *tablet* can improve access to other digital technologies, especially for those who have not yet mastered them or have no equipment at home, not to mention the ease of transportation: "The *tablet weighs less than a notebook",* some teachers point out.

I continued to administer the interviews according to the teachers' availability. In the daily life of the school, something stands out: the busy schedules of the teachers, some of whom even work in other schools in the capital. In order not to deviate too much from the planned timetable, I managed the opportunities that arose to carry out the interviews, bearing in mind the numerous commitments of the interviewees and the information about a possible teachers' strike.

I left it to the end of the process to interview the coordinator of the NTHE and, when I first made contact, she was receptive and willing to collaborate with the research. The center is housed in a municipal school located in the same neighborhood as the school under study. On the surface, the NTHE seems to be well set up, with an organized structure that is fully operational. When I arrived, I saw a group of teachers working on their computers, talking about something that involved the NTHE and the municipal schools.

The NTHE coordinator was enthusiastic about the work she does at the Center. However, she recently expressed concern about the permanence of the Center. SEMEC is going through a management crisis after the change of Education Secretary. The teachers even went on strike for a long time and were planning another strike. In a conversation with the coordinator of the Center, I found out that, with regard to the equipment provided by ProInfo, one of the major problems is the warranty period, maintenance and technical assistance. The coordinator revealed the following:

> "It's *beyond our competence. We are always dependent on the availability of technicians from the company contracted by the MEC. Another fact is the service system, each visit corresponds to the repair of only one piece of equipment. They belong to an outsourced company and earn per machine. So we have to go into the system and place an individual order, machine by machine, and that takes time.*"

Once the interviews had been carried out and I had the audio work, I began transcribing the speeches. This stage was one of the most significant of the work, as it was a very rich moment of information and inspiration. With each transcription, I discovered new things that would be important for my dissertation. In this process, I used the FFTranscriber 1.0 program to make the work more dynamic.

After all these steps, I was ready to start chapter 6, entitled "The paths of analysis and results". At first, I was a little dazed, because I had a lot of information and I needed to organize it according to the standards of a scientific paper, in all its rigor. At this point in the process, I resumed reading books and texts that theorize content analysis.

CHAPTER 6

ANALYSIS AND RESULTS

In this chapter I describe a comprehensive analysis of the educational reality of a municipal public school in Piauí, involving digital technologies, the computer lab acquired by the National Program for Technology in Education (ProInfo) and the pedagogical practices of the teachers of this Teaching Unit.

Here I present the way in which digital technologies have been used by teachers in their daily pedagogical practices and how they perceive these resources in the teaching process and, above all, their view of the computer lab acquired by ProInfo. All the methodological strategies were used to find ways of understanding my object of interest and analyzing it under the focus of the three thematic axes defined.

All the material collected and the texts produced at this stage are the result of visits and observations in the empirical field; extensive dialogues with teachers and the school unit's pedagogical coordinator; the coordinator of the Educational Technology Center of the Municipality of Teresina (NTHE) and, above all, interviews with semi-structured questions with previously selected teachers based on dialogues and the application of a questionnaire which is included in the annex to this dissertation. I would like to point out that, throughout the process, I have been honest, showing fidelity in the transcriptions of the interviews, in everything the interviewees said and throughout the investigation.

The paths we have traveled have been long, but full of surprises. In my view, this research, configured as a case study, reveals a scenario of hope and a lot of struggle, waged by education professionals who are raising a banner of improvement and development. It's about contributions to the success of the education system, contributions that start with the teaching process and work their way down to the subjectivity of each education professional. Although we are aware of the federal government's actions for digital inclusion in education, given the various programs created and implemented throughout the country, through this case study I was able to see that the majority of teachers also seek to acquire technical knowledge on their own initiative[17] [18] [19] . I reinforce this by transcribing what one teacher says, whose speech is representative of the thinking of many other teachers:

> "*At the beginning of the ProInfo program, we had a monitor who helped us in the lab. The first few times I took my students, I tried to learn from the monitor how to handle the machines, turn them on and off properly, connect them to the internet. I wanted to be independent in case he was absent. And so I learned to use these digital resources better*" (P1).

In this sense, the reality of the Teaching Unit reveals that some of the education professionals are seeking digital emancipation[20] based on their actions, experiences in practice and the use of digital resources to mediate teaching activities.

> "*[...] so we did an intensive course lasting two months, and I'll tell you: that's what opened the doors to the world of new technologies for me. I didn't know anything, not even how to turn on a computer. I did my specialization monograph by leaps and bounds when it came to typing. I had no idea. So it was this course that made me learn a lot of things, that really opened doors for me. It was a basic, quick course, but it made a lot of things easier for me. This qualification came about because of the interest of the school, the secretariat (SEMEC) and my own interest, but since then I haven't taken any more courses. I'm aware: what I know, what serves me, is in my own interest. The interest of researching, of discovering. I have children who sometimes help me when I have difficulties. And we learn*

17.
Self-education occurs when the individual participates independently, seeks to know, acquire skills on his own initiative and is aware of the objectives and results he wants to achieve.

20 The term "digital emancipation" was developed through the City of Knowledge Project, an initiative of the University of São Paulo's Institute for Advanced Studies by researcher Gilson Schwartz.

from the students too" (P1).

Even though they identify these particular efforts, the computer lab, which should be a space for promoting digital emancipation, at the school under study, the teachers say that this space is not working effectively, discouraging teaching practices. They recognize that the management of the program does not meet the objectives proposed in Decree 6.300, of December 12, 2007, a document that proposes a reformulation of the initial program[21] . The reformulation proposed by the Decree consists of the following:

> Art. 1 The National Educational Technology Program - ProInfo, executed within the scope of the Ministry of Education: will promote the pedagogical use of information and communication technologies in public basic education networks.
>
> Sole paragraph. The objectives of ProInfo are:
>
> I - promote the pedagogical use of information and communication technologies in basic education schools in urban and rural public education networks [...]

In 2002, the municipal school under study joined ProInfo. The program in the state of Piauí is the responsibility of the State and Municipal Secretariats. The Municipal Education Department (SEMEC) maintains the NTHE, a body created to manage digital inclusion actions in the municipality's schools. That same year, the computer lab was installed on the premises of the school. This space would be used for teachers' pedagogical practices using digital technologies. An agreement was reached with the schools to guarantee the necessary and appropriate conditions for the lab to function properly. These conditions include the electrical system, the laboratory's infrastructure, ranging from workbenches and an air-conditioned environment to other priorities.

As a result, training was organized by the NTHE for teachers wishing to enter the digital world. Here, I reinforce what Lévy (2000) points out when he says:

> It is not just a question [...] of using technologies at any cost, but of consciously and deliberately accompanying a change in civilization that profoundly questions the institutional forms, mentalities and culture of traditional educational systems and, above all, the roles of teacher and student (LÉVY, 2000, p. 172).

Following the same line of reasoning, Demo (2007) also emphasizes the fact that it is not enough to make computers or any other type of digital resource available, because in his view, the potential of pedagogical mediation with the use of digital technologies is based on:

> [...] the ability to use, for which *digital skills* are crucial. These are defined not only as the ability to operate computers and network connections, but above all as the ability to search for, select, process and apply information from multiple sources and, in particular, the ability to use information strategically to improve one's position in society. In this sense, access to the new media profoundly conditions opportunities to participate in many fields of society (DEMO, 2007, p. 14).

In schools, digital emancipation is established both through the use of technological resources and through interaction between subjects, forming networks to exchange knowledge in a collaborative process.

During the course of this research, *in* addition to the *on-site* observation process and informal dialog with the teachers, I did extensive work listening to and transcribing the interviews with previously selected participants. Once I had this material, I read in depth about the teachers' discourses and conceptions of digital technologies, ProInfo and their teaching practices. To help with the interviews and the organization of the information, I worked out the objectives in three thematic axes that I identified as fundamental to this research[22] . Based on this, I tried to analyze the relationships between teachers' pedagogical practices and digital technologies, mediating the teaching process.

In the process of surveying and analyzing the results, they were organized into thematic axes,

---

[21] Decree No. 6.300 of December 12, 2007, amends Ordinance No. 522, provides for the expansion of ProInfo, outlines objectives and responsibilities for the program, changing the former National Program for Information Technology in Education to the National Program for Educational Technology. Its foundation is also supported by some provisions of the LDB (Law No. 9.394/96), especially Art. 32, Item II, which aims to train

[22] See page 63 of this dissertation for the thematic axes.

which served as a parameter for identifying the dimensions and units of meaning[23] that involve the subsets of information that will make up the *corpus of* the analysis, written through relationships and/or connections established in the interviewees' conceptions. In this context, the units of meaning were considered by selection and similarity of concepts extracted from the interviewees' speeches, giving order, by importance or frequency, to the text of the dissertation, and underpinning the configuration of the educational reality of the school under study. All the data collected is linked to the interrelationship with the research topic and the objectives defined in the project.

Using a qualitative approach, in this research I try to gather information to understand the specific and particular nature of the object of interest, since the focus of my research is directed towards the conceptions of individuals. In this model, I intend to decode and understand reality as it is, as experienced by teachers based on the thoughts of those who work in the school environment. In this research, I try to value the most particular aspects of the subjects being investigated. In this respect, I understand the importance of the researcher's role in identifying the conceptions and meanings that teachers demand in relation to digital technologies, observing the facts that influence the dynamics of the contexts and reality of the subjects in the school environment.

I try to highlight the dimensions that reveal the teachers' conceptions of digital technologies, the meanings they give them in their teaching practices, as well as their representations of the computer lab acquired by ProInfo. I am also trying to highlight their conceptions of the qualification process for using these digital technologies in everyday school life.

6.1 Thematic axes and reflection based on the data collected

In this section, I try to record the results of the data collected using the methodological tools, as well as the statements made by all the people who took part in the research. The data is presented accurately and is interpreted throughout.

6.1.1 First axis: Teachers' experiences with digital technologies in pedagogical mediation.

6.1.1.1 Teachers and digital technologies

Based on this dimension, I observed that it is of fundamental importance for teachers to know and master digital technologies for the teaching process. At the school, almost all the teachers use them in their teaching practices, whether in class management or in planning and preparing content, a fact that is independent of the computer lab (ProInfo). They consider that there have been significant changes in the reality of the student and the educational environment. These changes include digital technologies and the teacher's didactic-pedagogical practice.

According to the teachers, the students are already introduced to the digital world. Some have already mastered it and own sophisticated equipment. They often bring various types of cell phones and *smartphones to* school. They use these resources for learning at school and, above all, for leisure and entertainment. All the interviewees were unanimous in saying that digital technologies are important for education. Even those who have not yet mastered these resources say that the school is moving towards a different scenario and that digital inclusion is already part of the school's objectives, although they emphasize the difficulties they experience in their daily educational life.

These are factors that go beyond the commitment and goodwill of education professionals, involving everything from the availability of training, an electricity system, favorable infrastructure conditions, adequate equipment, to the institutional commitment to maintain the conditions necessary for digital inclusion to really take place in schools, a common scenario in some localities in the public schools of the municipalities of Piauí. In this respect, Moran (2001, p. 51) states that for digital inclusion to be successful in the public education system, schools need to have guaranteed access and the process needs favorable operating conditions.

Among the various statements made by the teachers who took part in the interviews, I selected a few, identified in this research by units of meaning (US). For the teacher identified as P1:

> *"Digital technologies used as a teaching resource are extremely important for education. We have to keep up with these technological advances, not least because our students are already part of this context. It's a reality for them, so we teachers have to keep up with*

[23] In the context of this research, I understand a unit of meaning to be the consistent exposition of the subjective significance observed in the relationship between the accounts of the subjects surveyed.

these developments. In my own classes, I always try to make use of these devices: I use Datashow, the notebook, the students' cell phones".

Corroborating the above and finding a unity of meaning in the interviewees' speeches, I note the concept of P3, who emphasizes:

"I believe that working with digital technologies, especially with our audience, which is children and teenagers, attracts their attention more than a more traditional lesson" (P3).

P4, who also agrees with the changes in the way we teach and learn as a result of the introduction of digital technologies in education, says:

"I like digital technologies. I always try to innovate, bringing something different to the students. I research and look for new things on the internet. The internet is considered a new continent today; a continent of information and knowledge. I always try to research and find information on the internet, on websites and blogs, and bring it into the context of my classes" (P4).

This shows that teachers in general are interested in and committed to entering the digital world and are better prepared to use DT in the educational environment.

6.1.1. 2 Meanings attributed to dynamism

Teachers use different spaces to teach their classes when planning with digital technologies. In this respect, we identified from the teachers' experiences that the discourse involving technologies and the dynamism they recognize as a result of pedagogical mediation with the digital resources used in their classes is recurrent. Another important factor here comes from the fact that these technological resources promote interaction between students and the school environment.

During the research process, I chose to make a few visits to the empirical field in the run-up to the interviews. In informal conversations, some teachers report that when they use mobile technological resources, they generally choose spaces other than the traditional classroom. This gives the process dynamism and the opportunity to mobilize, factors that encourage student participation, awaken creativity and a certain amount of autonomy in the development of the work. As a result, these teachers plan lessons to be carried out using the students' cell phones and *smartphones*, configured in mobile technology with audio and video applications. In this regard, some teachers' reports highlight the importance of combining digital resources with pedagogical use in order to add value and encourage student participation in activities:

"[...] as a lesson plan, in the previous year (2012) we carried out activities using the students' cell phones. We made choreographies using the songs we already had on our cell phones, we made videos and shared them with the other students. We used other spaces in the school for these activities. We always try to use them. So I think it's very important". (P1)

I realize that digital technologies promote much more than dynamism when used to mediate pedagogical practices. From the teachers' accounts of their experiences, they have resources in their hands which, if used with "innovative competence" (DEMO, 2002), represent an opportunity to work with students on certain attributes involving citizenship, gender and diversity:

"In my activities, I try to encourage everyone to take part. The kids are shy and embarrassed at first, but as the class progresses, I work on the concepts of gender equality and diversity. I try to show them that dance is an art. They even used their cell phones with music and speakers. I didn't even know I could do that. We also filmed short videos with these devices. Then we put the work together and shared it with everyone at school. By then, they were all creating and making choreographic compositions. And what seemed to be an obstacle at first, the question of dance for the boys, later became something very interesting, because they even increased their work with the resources of the cell phones" (P1).

Another experience worth highlighting is that of a teacher who shows enthusiasm when talking about the project he has been developing for some time. He points out that he always carries out this activity with 6th graders, who are just starting out at school. As well as taking advantage of the students' equipment, cell phones and *smartphones,* he does an activity similar to a walk-through lesson, but with the recording of images and speeches by interlocutors. According to P2:

> *"I decided that we would work on producing texts, news and reports using audio and video from the students' cell phones. What used to be a problem became an excellent teaching resource and an incentive for them. After producing the material, we had a moment to present the video to the other students at a cultural fair organized by the school. They saw themselves in the video as reporters, interviewers, at the same time as spectators.*
>
> *In this production, we worked on raising awareness, recognizing the spaces in the school, from the library, the playground, the classroom to the toilets. Then they realize how ugly the school is, all scratched up, plundered, the toilets scratched, dirty. Some students urinate outside the toilet. We worked on the issue of education and valuing space and respect for public property. We took the opportunity for them to draw a picture of the school and for them to see themselves in this environment. As well as producing the text, they come to these conclusions which involve citizenship and responsibility"* (P2).

This teacher, when planning and creating activities with technological resources, demonstrates that there is a need for mastery of the technique and the awareness of the educator to develop strategies that prioritize changes in behavior and worldview, which I believe is also the responsibility of the school.

In this way, I'm trying to identify a relationship here with what Freire proposes, when he says that the educator's task, then, is to problematize the content he mediates with the students and not to dissertate on it, to give it, to extend it, to deliver it, finished. In this act of problematizing the students, he is also problematized (FREIRE, 1992, p.81).

The account of this experience is in line with my questions about the way in which technological resources are being used by teachers in their teaching practices. I can see from the teachers' conceptions that teaching with technologies, depending on how they are used, can add value to teaching. I also realize that this management of digital technologies generates a form of self-knowledge and awareness on the part of the student, as a participant in the teaching and learning process.

According to Demo (2002), by acquiring innovative skills, teachers can promote successful teaching practices, especially those that include the use of digital technologies. In this sense, technological resources, mediating teachers' practices, can generate emancipatory knowledge, as some concepts and values promoting human evolution are being worked on, such as respect for others, ethics, citizenship, political participation, among others:

> *"Regarding the area that many people don't value, which is the area around the school, which generates a lot of problems, violence, cell phone thefts, fights, confusion, the students filmed everything from the entrance, then the break time, leaving the school and even how these students behave around the school. Because we received a lot of complaints from people living in the neighborhood. Neighbors complain about fights, ringing the doorbells of neighboring houses. The result of this work is always very good and positive"* (P2).

In the dynamics of these activities, with collaborative approaches, I identify determining factors that promote digital inclusion and the emancipation of subjects. Students are given a greater opportunity to exercise their autonomy and a certain degree of responsibility in carrying out the work proposed by the teacher. In this interaction, the student moves from being passive to being an active and creative manager of learning. In this way, I can identify a student who is also a builder of knowledge. According to the National Curriculum Parameters (PCN): "The school has an important role to play in society, teaching students to relate critically to the universe of information they have access to in their daily lives" (BRASIL, 1998, p. 139).

Regarding these experiences with digital technologies, Demo (2007) makes important points, considering the use of digital resources as a significant part of the self-learning process, when provided by teachers when developing pedagogical activities. As a result, students have the opportunity to evaluate their experiences, the results and their reflections on the knowledge that has been built up in groups, in short, their cognitive development as subjects of their educational history.[24]

Pondering the fact that collaborative activities do not exhaust the possibilities for learning, the experiences reported by the teachers make me believe that students, by participating in these dynamics, achieve relevant and expressive learning. Schwartz (2008) explains that nowadays it is no longer enough to be digitally included. According to his reflection on the subject, it is now of fundamental importance to be and feel digitally emancipated. For the author, an emancipated person goes beyond the passive use of digital technologies, information and communication technologies, mainly by using technological resources consciously, critically and autonomously. This tends to make the results obtained in traditional digital inclusion projects more effective.

Emphasizing these facts, I would like to highlight Almeida's view of school as a space where teachers are responsible for providing an environment that: "stimulates thinking, that challenges the student to learn and construct knowledge individually or in partnership with colleagues, which fosters self-esteem, sense-critical development and responsible freedom" (ALMEIDA, 1999, p. 21).

In this sense, I believe that the school, much more than training students to use digital technologies, is a place for training subjects for the world, prepared to reflect critically, deal with new skills and know how to live in a constant process of change, in line with Paulo Freire's thinking, for whom education, starting from the student's reality, must be configured as emancipatory, transforming minds and attitudes.

6.1.1.3 Digital technologies and classroom management

For teachers, digital technologies centered on computers and the *internet* represent significant changes in the way they research and develop content for their classes: *"We no longer have just the textbook,"* says one teacher. The *internet* represents a space of great opportunity. According to the teachers, when they plan their lessons, they always make use of digital resources, either in the production and presentation of content with appropriate *software, or* in the search for content that addresses the topic they are going to develop with the students, or even videos and texts for reading and group discussion. In this respect, I would like to transcribe some of the interviewees' significant comments:

> *"I use the internet to research and prepare content for lessons. In physical education, there are sports rules that are always changing, always evolving. We need to keep up with these changes. I can't keep working on obsolete content with the students. I have to keep up and I can't afford to buy books with new rules all the time. If I have the chance to buy them on the internet, then I go there, download them, save them and pass them on to the students. I put it on the datashow, present it and discuss it with them. When I take the students to the laboratory, I already work on something that I can actually network, so that they can see something specific to physical education at the same time, such as images of a court, short videos. I put together a collection of short videos of rhythmic gymnastics, aerobics, on the Internet for them to view. So that, through visual communication, they can learn much more about knowledge. There is an infinite number of things we can search for on the internet. As well as looking for new publications, the work of researchers in the field who are always innovating. These resources are an inexhaustible source of information"* (P1).
>
> *"I have internet, a laptop and a computer at home. So I like to search the internet for texts, videos, to make slides with class content, take photos, images. I'm always researching or writing texts to share with the students. So I'm always bringing something interesting that I find by searching the internet. For example, youtube videos, an internet documentary. There's a website called "Porta Curta Petrobrás" video shot which I use to record interesting, short videos of between 5' and 10' to show in the classroom and discuss, ask questions that involve the theme of the videos and work with the students"* (P4).
>
> "Digital technologies [...] *They help a lot. I have a netbook and without it I'd be lost. On the internet, sometimes you find something interesting on a website that you can apply in*

[24] See the quote from Demo (2007) on page 76 of this dissertation.

the classroom or you find an experience that you can adapt to the content you're going to apply or you find a blog or website with an experience that someone has already done and it worked. All of this motivates research and the search for new experiences, because we don't learn from nothing. [...] So, the internet, when it comes to researching and preparing teaching materials, these resources certainly help a lot. You get out of the textbook a bit and see what other teachers have done elsewhere, even in other areas, which in a way encourages us to look for something new on the net. All this either gives us ideas or the opportunity to adapt to our classroom reality. Something done in another city can't always work here. Even here at school, each classroom has its own specificities. Sometimes you plan to socialize some content and in one classroom it works, in another it's a bit difficult, it doesn't work so well, even though they're in the same grade, but the clientele is different. So it's important to be aware of these diversities and, based on that, adapt good examples to our reality" (P2).

As you can see, the various reports transcribed here show the presence of digital technologies in the activities carried out by teachers; they are pedagogical tasks that precede classroom management. In conversations with the research participants, I found that the vast majority of them use the *internet* and *software* to plan and produce lesson plans. It is often said that in recent decades education has undergone significant transformations. For some, there is no going back. Students are already familiar with these technologies and schools need to invest more in infrastructure, equipment and programs to better adapt to this new educational profile. I also found that all the teachers have computers and a network connection system in their homes, but not all of them are proficient in using these resources. However, the vast majority use them in their teaching practices, such as research activities, planning and preparing content for lessons. They recognize the importance of these digital technologies for education and say that they are preponderant today for teaching and learning.

In this scenario, I would like to highlight the account of a teacher who says she has no command of these resources, admitting that this fact makes her feel embarrassed at school.

"My classes are expository and I also use the textbook, pictures and maps. Even if I don't use a computer or the internet, I always try to innovate in the way I run my classes. I look for different ways to present the content to my students. I don't have a good command of digital resources and, as there is currently no one to help us in the laboratory, I try not to take the students there. I think that would be too difficult for me. I'd like to show them the maps, work with them on internet searches, but I don't feel very comfortable because I don't know how to use those machines. I think it's embarrassing for the teacher to realize that the students often know more than he does" (P5).

Based on the teacher's comments, I can see that there are still teachers who don't use digital resources and don't usually go to the computer lab, because they feel insecure about these advances. However, they recognize the importance of technological resources and the possibilities of using them in teaching practices.

In this sense, technological developments are placing new demands on the teacher's profile. In this respect, I recognize that many teachers still use all or part of the traditional model of teaching, show difficulties in handling these technological resources, as well as in incorporating them into their teaching practices. All of this reflects a movement towards change, causing insecurity for teachers as "digital immigrants" in Lévy's (1999) conception. The need for digital training for education professionals is therefore evident, and this issue will be addressed here in the **second thematic axis**. This corresponds to training in technology and its impact on teaching practice, from which dimensions have been selected.

It's worth noting that the use of digital technologies in the classroom can be considered a first advance in the teaching process. Even so, the traditional approach has not been supplanted, since the logic of transmitting content persists, today only with greater dynamism, through the use of more modern audio, video, etc. resources, which does not constitute an aspect of emancipatory education, and it is therefore necessary to abandon the pedagogical mediation of a banking nature in Freire's sense.

In this sense, through this research, I was able to see a manifestation of emancipatory education through reports of the experiences of teachers who use technological resources, especially those that their students deal with on a daily basis, giving these resources a condition that goes beyond mere technical objects, by also treating them as a means of putting into practice the concept of citizenship, social responsibility and political participation[25] .

6.2.1 Second axis: Training process for the use of digital technologies in education.

6.2.1.1 Offer of courses and availability of access

The Teresina Educational Technology Center (NTHE) is the body responsible for training teachers in Teresina's municipal public schools. Through the NTHE, teachers have access to training in digital technologies for pedagogical use in the teaching process. The NTHE began its activities in 1999, when a partnership was signed between the Ministry of Education (MEC) and the Teresina Municipal Education Department (SEMEC). The selection criteria, according to the coordinator of the Center:

> "*is due to the favorable conditions of the school in terms of infrastructure for the installation of the laboratory, an air-conditioned room, an adequate electrical system and a security system to protect the equipment.*"

Following the implementation of the computer lab at the school, a group of teachers from the NTHE took part in a specialization course in education and technology offered by the MEC. After the training, they began to work as multipliers in the municipality. Since then, the NTHE has been offering courses to train public school teachers to use digital technologies in their teaching practices. In addition to the courses suggested by the MEC, the NTHE creates its own courses and workshops to meet the demands of schools in the municipalities. According to the NTHE coordinator, the initiative and interest for these training courses must come from the school that has the ProInfo laboratory. However, the schools served by the program always receive information and invitations to the workshops and courses offered by the NTHE.

In this regard, during the time I was at the school, I saw an invitation from the NTHE on the wall of the teachers' room. A call for teachers to take part in training. When I asked them about their participation in these courses and workshops offered by the center, the interviewees complained about the way these trainings were managed. At the school, I found that some teachers always take part, while others talk about the difficulties. The latter complained about not being excused from classes during the training periods. Here is the testimony of one teacher:

> "*I'd like to do some computer training. I even see the NTHE offering courses for teachers from time to time, but I've never been able to do it. There's no time for that. We're not exempt from classes. So, with so many activities to do, there's no time left to train in this area. But under other conditions I would do it. I see it as a necessity today. I also know that everyone has to be interested. An interest in learning, in changing.*" (P5).

In the course of the interview, the teacher emphasizes factors that, in a way, interfere with the process of digital inclusion in education.

> "*I know that these technological resources are important for teaching, but I think there is a lack of incentive on the part of the school and the Municipal Education Department (SEMEC). I think strategies or ways of involving teachers in this training need to be devised.*" (P5)

When asked about the greatest difficulties encountered at school in including digital technologies in their day-to-day teaching, some of the interviewees responded:

> "*It's the question of incentives. We need time for training. If the teacher wants to learn how to use these new resources, he has to make time for it, because the school doesn't make that time available. They have to find the time if they want to get up to speed*". (P2)
>
> "I've *never taken any courses. What I know I learned on my own. I've heard about various courses, but I've never been interested due to lack of time. The ideal thing would be to make the teacher redundant or to make the course available in the school itself, since it*

---

[25] See, for example, teacher P2's experience report on page 83.

has the ProInfo laboratory" (P3).

In general, teachers in the municipal education network are assigned to 40-hour shifts. Another factor that prevents them from taking part in training is the number of teaching units to which they belong. In the municipality, the school system is organized by zone. Several schools are integrated into one zone of the capital. In this organizational design, some teachers teach in more than one school. Here's one teacher's account of this:

> *" I always use digital resources at home when I'm preparing lessons, doing research to plan activities, exam questions, reading texts, suggesting books, films and videos. But I haven't yet been able to use them in the classroom. It's not that the school doesn't have these resources, it has a data show, a notebook, but I haven't organized myself to use this technological equipment at school, in the classroom. I think with me it's because of the size of the classes, the rush, the very strict timetable. All this demands time for you to carry out these tasks in the classroom. Because of all this, I don't use it and we let it go, time goes by. I'm a teacher in another school system and you end up with a lot of demands. I try to make classes very participative, but without using digital technologies"* (P3).

I recognize that, with so many commitments, it is almost impossible for teachers to invest in training or lesson planning that includes the use of technology. In other words, while on the one hand the government offers strategies for digital inclusion through its programs and public policies, on the other hand the local reality is configured in situations that are incongruous with the objectives of these initiatives. All that remains is for teachers to find the time to invest in their professional development, even if this means giving up what time they have left for leisure or entertainment.

When I interviewed the coordinator of the NTHE, she said that one of the biggest difficulties that interferes with teachers taking part in the courses and workshops is the fact that city halls and municipal education departments insist on not releasing teachers during this period. With this recurring problem in mind since the beginning of the program and the training of the NTHE, training courses are now offered with part of the course online and part in person. This confirms the teachers' reports. In this sense, I base my questions on the words of the coordinator of the Technology Center:

> *"What is lacking is better monitoring and greater pressure from the MEC in relation to the counterpart of the municipalities and the state, which is not always forthcoming. For example, when the municipality joins the program, it has to ensure that the teacher has this qualification and, often, the mayor doesn't release the teachers for this. Interested teachers have to find the time to do this qualification on their own. Most of our courses today are semi-presential to see if we can reach a greater number of our clients. It was no longer possible to continue under these conditions. We have one meeting a week and the rest is distance learning".*

From the coordinator's account, I can see that in general the federal government's public policy programs, in terms of their formations, objectives and goals, are important for the population. However, there is a lack of greater monitoring by prosecutors of the conditions and results of these programs. In most cases, this whole structure has its bottlenecks at the end of the process, in the locality for which the program was designed and made available. In the case of the federal government's digital inclusion programs, several factors interfere with the conditions in which they operate.

Even with so many problems identified in the course of the research, it became clear that change also depends on the interest of education professionals in wanting to participate and evolve. Some of the participants in the survey were clearly enthusiastic about digital technologies and their possibilities.

Some facts are recounted by the NTHE coordinator when she was interviewed. In the conversations we had, her enthusiasm and persistence in fighting for changes in education were clear.

> *"The LDB says that the state and the municipality have to guarantee this training for teachers, but we know that it all depends on whether or not those at the top value this training. Sometimes they (teachers) start and then have to give up because they can't reconcile their time at school with the training offered by the NTHE. That's why we invest*

a lot in the workshops, because we do them in the school itself. We teach two hours one day and then come back and teach another two hours. And so we continue to qualify, train and support the teachers in the schools. We're not giving up on the program. The progress we've made since the beginning has been enormous. We started with 120-hour courses and 12 schools with laboratories. Then we went up to 17 schools. Then other workshops came along. During the year, we used to hold 4 to 5 workshops. Today we have more workshops and a great diversity of themes."

It should be noted that the legal framework exists, but that in order to implement what it says, movement is needed, not just from government officials, but from the subjects, groups and the school community concerned. In this respect, Moran (2001) observes that: "Society needs to have as a political project the search for ways to reduce the distance that separates those who can and those who cannot afford access to information" (MORAN, 2001, p.51).

Qualifying teachers to use digital technologies has proved to be a major challenge since the process of computerizing primary schools in Teresina's municipal education network. With the implementation of computer labs in schools, a number of needs have arisen that involve teaching tasks, such as: incorporating digital technologies into the teaching process and changing the vision and attitude of teaching. Reflecting on these factors, all the teachers interviewed highlighted the importance of ProInfo and the work of the NTHE. Regarding teacher training in digital technologies, they admit that it needs to be continuous, pointing to the idea of constantly seeking knowledge and acquiring skills. Lévy points out that qualifying teachers to use digital technologies is an imperative in a school that is part of cyberculture (LÉVY, 1999).

The informal conversations I had with some teachers revealed a concern about continuing training that takes into account the transformations and changes in the educational scenario, such as the introduction of digital technologies in education, which undoubtedly lead to significant changes in teaching practices. Some question the way in which this qualification process managed by SEMEC and the NTHE is being conducted. In this respect, they admit to managing their qualifications as far as possible. When asked about training courses for the pedagogical use of computers and whether this initiative comes from the Municipal Education Department (SEMEC), one statement reveals the situation that municipal education is currently going through, given the constant strike demonstrations. The most common reasons for these demands are the Career Plan and the low salaries of education professionals.

"If you're encouraged to do this, to train with technology, to take courses in the area and you get a return, then we're motivated to take part, because we know we're going to improve our salaries. But if you don't get any feedback, you'll only do it on your own initiative to improve your knowledge. That's because you want to learn, you want to apply new things to your teaching practice. Ideally, this should have a change of class in the Career Plan, it should count towards a change of level. You're not going to do it just for the sake of doing it. If it had a greater return, it would be much better. We always have an objective for the future. (P2)

Even though the teacher was unmotivated by the system, he revealed that he had already signed up for a course that the NTHE was offering. The course would be *Phliks* in math and Portuguese. He reveals that they are games similar to *video games*, but with pedagogical *software* adapted to teaching. They are in the form of challenges and students go through stages until they have overcome all the challenges.

"At the end of this semester (2013), a program acquired by the municipal education network in partnership with the Ayrton Senna Institute was presented to teachers. It's called the plinks digital platform, aimed at using games in early childhood education. It's a kind of educational game, a challenge for the students. We, the Portuguese language and mathematics teachers, are being registered to access the qualification and use the program at school. We were invited by the NTHE to be trained to use this platform. It's all very new at the moment. We're just starting to register, but I don't even know if we'll be able to use it this year (2013). We don't know how it will be used here. Not least because

the laboratory isn't working properly at the school. Some of the machines have problems and the internet is very slow. But this program requires the use of the computer lab (P3).

This NTHE initiative, reported by these teachers, demonstrates that the program enables the use of digital technologies and *software* that can be used to enhance teaching. These are actions that stand out and motivate teachers. In the light of what has been said, I can see that there is a desire among the teachers to innovate, to do things differently, but they show insecurity about taking the course.

Although there are already many training courses for the use of digital technologies in education, I understand that technological resources alone will not be able to transform traditional pedagogical practices into innovative ones, oriented towards the emancipation of subjects. Teacher training must work on development that goes beyond technical skills. It involves pedagogical skills oriented towards the reconstruction of knowledge or "reconstructionism" (DEMO, 2001, p. 28). With regard to teaching and training for the use of technological resources in pedagogical practices, the desired changes involve a change in the teacher's profile, the enhancement of skills and new knowledge, a fact that I believe represents a gap in the initial training of the vast majority of teachers.

After reading books, researching academic papers, articles, dissertations and theses on the subject of technology and education and, in particular, this period of research in the empirical field, I realize that digital emancipation doesn't just happen by making computer labs available in schools. This action is an obvious and fundamental initial step, but not a satisfactory one. Other studies have already pointed this out.

6.3.1 Third axis: Teachers' conceptions of the computer lab (ProInfo)

In order to substantiate the reality of ProInfo in the school under study, I present descriptions involving the conceptions of the teachers in this teaching unit based on what was identified in the **third thematic axis,** which seeks to obtain information from the subjects about the computer lab acquired by ProInfo and the relationship between the use of teachers in their teaching practices in everyday school life. In this axis, I identified the following dimensions:

6.3.1.1 Effectiveness of the computer lab

In this section, I propose to report on the teachers' conceptions of the computer lab purchased by ProInfo. After talking to some teachers during the field observation period, I recognize that the reality surrounding the conditions for using this educational space is in need of reconfiguration, from the infrastructure of the place to the conditions in which the equipment is located. The neuralgic vein surrounding ProInfo goes beyond quantitative aspects, extending to qualitative ones. The presence of a computer lab in the school generates controversy among teachers regarding the ways in which computers are used, the teaching process and, above all, the conditions in which the lab is currently located in the school.

According to Chaves (1998), quoted by Silva (2010), the presence of computers in schools is only justified if they can help the school to perform its functions better, if there is a commitment on the part of managers to introduce computers, effectiveness in the process of managing this technology and if they represent an important pedagogical resource for teachers.

In order to meet these conditions imposed by the new educational reality, the school is expected to determine appropriation strategies for the use of digital technologies in accordance with its pedagogical project. In this case, technological resources become an effective part of teaching, helping teachers and enhancing teaching practices.

In the interviews with the teachers, when they were asked about their opinions of ProInfo, their facial expressions and words showed discouragement and disappointment. Even though they admit that they have been encouraged to become more involved with digital technologies since the lab was set up, they say that over the last 11 years, ProInfo has undergone some changes which are reflected in the way the lab works:

"When it (the laboratory) was first installed, there was a contribution. If this program hadn't come, I don't think much would have changed in teaching practice. The fault lies in monitoring the process of permanence and maintenance. The project was launched and implemented, the laboratory was installed. As time went by, the project fell by the wayside.

When it started, there was good participation and the teachers were very excited. There was even a time when there was an incentive, a facility to buy
I think a lot of teachers bought computers at that time. The city gave 50% of the cost and the teacher paid the other 50% in installments. I think many teachers bought computers at that time. (P1)

"Here at school, the computer lab is insufficient in terms of the number of computers, and the internet is not of good quality. So, in a way, this makes it difficult to do research or develop an activity in the lab. Here at the school, there are many barriers because the laboratory is small, although it is well equipped with air conditioning, but the quality of the internet and the equipment is very poor. In the lab today, there is no one to support the teachers. If you want to go, it's at your own risk. I'm still able to do something because I'm able to control the situation, turn on the equipment and monitor the activities, and that's when I decide to do something in the lab" (P4).

On this issue, all the teachers who took part in the interviews expressed their dissatisfaction with the fact that the laboratory does not operate under favorable conditions. There are recurring complaints specifically about outdated equipment, a network connection system that is always very slow, and a much smaller quantity of equipment compared to the number of students in a class at the school:

"We suffer from a lack of maintenance, technical support for the machines, the internet is a struggle. We always encounter some problem. It's a struggle to resolve. There's also the issue of monitors, which we really need. There are more than 20 computers, almost 30, but only half of them work. So we need this technical support. The program needs to improve in this respect. At the beginning of the ProInfo program, there was a lot of competition between teachers to get access to the computer lab. Some teachers even presented projects. The Portuguese teachers made newspapers and developed some more projects to be developed in the lab. But this year (2013), with the absence of these monitors, we realized that the lab was somewhat underused" (P1).

When asked about PROINFO and the possibility of changes in traditional teaching practice, teacher P1 was emphatic in stating that there have been changes and that these new practices enhance learning. According to P1:

"The history of information technology in education was an important moment. The school needs it, we need students doing research, we need students having access to these digital technological resources. The student has something more, more motivation, more interest in building this teaching and learning process. The laboratory has always been used a lot, it's something important within the school. Students go to research when they need to do some work, it's an access that the student has. If students don't have a computer at home or in the surrounding area, they have to come to school and the school has to welcome them. So it's important. But it would also be important if it worked 100%, with all the conditions" (P1).

It is clear from the interviewees' statements that digital technologies are extremely important in education. They recognize that students feel encouraged to take part in activities involving these digital resources. However, the computer lab has problems that generate conflicts in the school. When asked what conditions the lab offers for teachers to carry out their lessons as planned, there were many complaints, as one teacher put it:

"If the lab was fully functional, this space would certainly be hotly contested by the teachers here at school. The problem is that you go once, twice, three times and difficulties arise. We get discouraged. I'll say it again: we need to make the lab more dynamic, we need someone there permanently to support the teacher. It's not enough just to put machines there. There has to be someone there to help put these machines to work. If ProInfo provided all the conditions, it would be different. The laboratory would certainly be used a lot by everyone. Today, it is underused. It's mostly used for Mais Educaçao. With regard to energy, the school works, but the internet is very slow. When there are 15

machines working, the system is always very slow" (P2).

"When I went to talk to someone at the school about the Plinks platform, I asked if I could register the students. He said it couldn't be done here at the school because the computers and internet weren't up to date. It had to be done elsewhere. That makes everything difficult. How can I want students to access this platform if I can't keep up with them? Ideally, they should be able to access it right here at school and the teacher should accompany them" (P3).

I would like to point out that the two teachers said they were enrolled in a training course to be offered by the NTHE and that they were familiar with the *Plinks* Platform. However, while they are eager to learn something new, they are also afraid of the difficulties of using it at school.

6.3.1.2 Monitoring and boosting teaching practice

The aim of the NTHE's monitoring program was to create opportunities for schools to work with technological resources after the computer lab was set up. This proposal aims to make the pedagogical use of digital technologies more dynamic, promoting the improvement of teaching in the school context: "In this way, teachers can experience different working practices and strategies with the presence of a monitor who helps them explore and use technological resources in pedagogical actions". (CASTRO, 2010 p. 60).

The availability of a monitor in the schools that joined ProInfo was based on the difficulties identified by the NTHE. These include the teachers' lack of technical mastery of the equipment made available by the Program; difficulty in adapting content to the *software* and applications made available by ProInfo, insecurity in handling technological resources, among other difficulties identified in the interviews with the teachers who took part in this research.

With regard to monitoring, all the teachers said that when the computer lab was first set up, the NTHE provided a monitor to help the teachers. In the course of the conversations, they showed some concern about the absence of this person, who they consider to be of fundamental importance in the lab. In this research, I identified some teachers who are not yet familiar with digital technologies, but admit the importance of these resources in teaching practices. And they point out that when there was a monitor to help, they felt more encouraged to plan lessons and teach them in the computer lab. It can be seen that, in the teachers' view, the professional monitor does not replace the important role of the teacher, but can provide support to users of the laboratory, generate digital integration, training in the use of technologies and promote digital inclusion. (CASTRO, 2010 p.61)

According to the coordinator of the NTHE, the multiplier teachers gave the monitors all the guidance they needed according to the demands of the schools, as well as doing follow-up work, with the aim of contributing to a better performance of the monitorships.

When asked about the representation of monitoring and pedagogical work, the majority of teachers emphasized that when there was this professional in the school, the performance of activities in the laboratory was more productive:

"If the lab had a monitor it would be more interesting. There is no longer this person to support us. The monitor can help the teacher a lot in carrying out the pedagogical work in the lab. The teacher has to go to the lab and have to turn all the computers on and off, get everything working, connected to the internet, and in the end we waste a lot of time. By the time you see it, class time has passed. We really need a person, a monitor, to help us control and keep the equipment running. If, when we got there, the machines were all ready to work, it would be very different. All this means that the teacher is unable to manage these problems. The issue of maintaining the equipment is one of the biggest problems" (P2).

However, one of the teachers interviewed said that even though he was aware of all the difficulties in the computer lab, he liked to take his students there to experience the technologies. He recognizes that many of the students at the school can't afford to go to the *internet* and don't even have a computer at home. Therefore, in P4's view, the school is still a place where students can have the opportunity to come into contact with these technological resources:

"I use the lab regardless of whether I have one or not. Even if I don't have a monitor, I go

to the lab. I like to give students the opportunity to experience a different environment. I like to use it to work with images and videos. I recognize that, without the monitor, it's more difficult, but I negotiate with them (students), controlling them so that they don't go to other sites, blogs, Facebook. But it always works out. I reiterate that I like to give students the opportunity to see different things and have contact with these technologies" (P4).

The teacher concludes his speech by adding:

"In a way, the students are getting to know these technological resources. But if the ProInfo lab were more up-to-date, with a monitor to help, the internet working properly and at an adequate speed, it would certainly be an opportunity to get them to know and learn about technology. To feel what a computer is and how they can use it to their advantage" (P4).

One fact that was evident in the teachers' statements is that, in their view, the monitor's work is more geared towards solving the technical problems of maintaining the computers in terms of their operation, such as the process of turning the machines on and off and also keeping them connected to the *web*. However, the monitoring team is prepared to go beyond these technical activities. According to the NTHE coordinator, at the beginning of the program, the monitors underwent training by multiplier teachers prepared by the MEC in partnership with some Brazilian universities. This training goes beyond mastering the handling of equipment; it extends to mastering the use of educational *software*, accessing the *Internet,* searching websites with educational content, among other skills and abilities.

In terms of potential, technological resources can certainly bring about changes in the educational system, particularly in teaching practices. On the other hand, in terms of limits, there are still many difficulties in managing and using them.

and how to incorporate them into the teaching and learning process, so that students can master these tools and, as a result, build knowledge individually and collectively. Only in this way will students learn effectively with technologies, in true digital emancipation, which seems to be happening very little in practice. In order to do this, it is also necessary to temper the fascination with digital technologies through work focused, in this case, on their potential in terms of pedagogical mediation. In this respect, I would point out that the use of technological resources as educational tools implies solving problems in the school context at the level of the internal and external community, understanding everyday educational life and acting to transform it. In this sense, there is still a lot to be researched, experimented with and discovered, considering the potential and limits in the use of technological devices in today's context, especially in the educational sphere, a question that I leave here as a challenge that could be explored in greater depth in future studies.

FINAL THOUGHTS

At the heart of this research are issues involving education, digital technologies and teaching practices. The purpose of the work was to carry out a comprehensive analysis of teachers' conceptions of digital technologies in education; the conditions in which these resources are used in teaching practices as a means of mediating the teaching process; as well as to find out how teachers view the computer lab purchased by ProInfo, which is available in the Teaching Unit for carrying out teaching activities using computers and the *internet.*

The reflections set out here are imbricated in the results obtained in the actions and activities developed during the research period, whether through reading books by authors on the subject, such as Adams (2010), Castells (2004), Lévy (1999), Tedesco (2004), Schwartz (2008), Oliveira (1999, 2001), Lopes (2010), Schlemmer (2011), Moran (2004), Demo (2002, 2007), Valente (1999), Freire (1987, 1996) and in official documents, journal articles, dissertations and theses; or in activities carried out in the empirical field.

On site, I used methodological resources such as observations, informal dialogues recorded in a field diary and transcriptions which are included in this dissertation. In order to deepen the investigation, I used interviews with semi-structured questions, supported by a script previously organized by thematic axes. These questions were applied to subjects selected through a questionnaire and/or dialogues during the visit to the school under study. In this way, I believe that, by reflecting on the data collected, it was possible to identify a contribution to the problem raised.

In summary, I can say that the teachers at the municipal school in Teresina, the empirical field of this research, recognize the importance of digital technologies for education. They admit that these resources, if well managed, enhance teaching practices and allow teachers to create and develop more dynamic, creative and innovative lessons. They recognize that these attributes are of great value for involving students in lessons, because, according to the teachers interviewed, students are already inserted in the technological world. Therefore, the only thing left for the teacher to do is to find ways of integrating them into the digital world.

I observed that, at the school, a small number of teachers are still not sufficiently prepared to work with digital technologies, to use them to mediate their teaching practice and to set teaching objectives with their use. In this respect, I recognize that the teachers who use computers the most in their teaching activities are those who have some knowledge of the area and/or have already taken a training course offered by the Educational Technology Center (NTHE). However, their great difficulty in reconciling their teaching activities with their participation in the training courses for the pedagogical use of digital technologies offered by the NTHE was evident. For some, the effort is often personal and private. However, they admit that this situation is almost always a challenge, as it usually comes up against the clash of timetables in their classes at school. In this regard, they acknowledge that there is a gap in the federal government's programs involving school management and the municipal government of Piauí. If, on the one hand, school managers are not responsible for releasing teachers to take part in training during working hours, on the other hand, it is clear that the majority of education managers in the municipality do not comply with the agreement signed with the MEC.

In other words, the real situation is incongruous because, on the one hand, the federal government creates and implements social programs aimed at digital inclusion in education. On the other hand, at the top end of the system, managers don't create mechanisms to make it easier for teachers to take part in education and digital technology courses. Faced with this scenario, I recognize that the Ministry of Education urgently needs to create strategies for returning/valuing these investments, as well as creating actions that can improve the proposal for digital inclusion programs in Brazilian education.

According to the research, it became clear that, through digital technologies, it is possible to plan actions and create activities in educational contexts that point the way towards the digital emancipation of subjects. In this regard, I would like to record in this section the experiences reported by the teachers identified here by the pseudonyms P1 and P2, who, in the interviews, shared some activities that go beyond the development of technical skills, but extend to working to raise students' awareness of values that involve a sense of citizenship and social responsibility. In their accounts of their experiences, these teachers recognize that, by appropriating digital technologies, they can use these resources to do much more than work on digital inclusion in a technical sense, by working mainly on qualities that point towards the emancipation of the subjects. In my opinion, this is what these teachers did when they developed tasks that included some important values for society. The aforementioned teachers chose to work with technological resources that are already part of the students' universe: cell phones and *smartphones*. They point out that the choice of technical objects is the result of wanting to work with the students' skills, since they have mastered the technologies in these devices. The teachers point out that, through these activities, the students can see that the technical object (cell phone), which was previously used more as a means of communication and entertainment, can now help them with school learning activities. Teacher P1 tried to develop dance activities with the students using the music they heard on their cell phones. However, the difference was that the activity was not a simple choreography; on the contrary, the teacher introduced the notion of groups, gender equality and diversity into the educational context.

Overall, I can see that the way teachers use digital resources in an educational context makes all the difference. The example cited above shows that, based on this activity, the lesson was configured as an enhancer of the development of competences, skills and knowledge shared in the school environment. This perspective indicates the possibility of emancipatory pedagogical mediation that is linked to the development of critical, creative thinking with an investigative attitude. But this practice is still restricted to a minority of teachers at this school. The majority of teachers are based on the idea of digital inclusion in the sense of learning to use technology.

The results of the research show that digital technologies are present in the pedagogical practices of teachers, who, for the most part, use these resources in activities during the teaching process, whether in the planning that precedes the lesson, or in practical activities developed with students at school. The reality of digital inclusion in pedagogical practices is also the result of teachers' efforts, through individual practical experiences or the training offered by the Teresina Educational Technology Center (NTHE). All the teachers are aware of the urgency of appropriating these innovations in order to adapt to the new student profile, which usually demonstrates skill in handling digital technologies.

Based on the teachers' thoughts about the computer lab, expressed in the form of questions and/or complaints, I found that, for some of the teachers, the lab is still an important space in the school, but they admit that it is fundamentally important to resize the management of ProInfo in order to meet local needs.

According to the teachers, the computer lab, created to be a space for promoting digital inclusion in schools, is not working according to the objectives set out in Decree No. 6.300, of December 12, 2007, which deals with ProInfo. This being the case, I believe that public policies are needed that can guarantee resources to continue the qualification of education professionals, through the actions of existing programs, so that they meet the objectives and goals set out in official federal government documents.

In the certainty that digital technologies can contribute to social justice, this research takes as its main issue the fact that the ability to access and acquire new knowledge through these technologies is an important factor for social inclusion from an emancipatory perspective, i.e. leading subjects to a process of autonomy, creativity, social and political participation, ethical posture and socio-environmental co-responsibility.

I found that digital emancipation doesn't just happen with the installation of computer labs in schools. This action is a fundamental first step, but not a satisfactory one, as the results of this research point out, when they identify the problem of the laboratory, which ranges from infrastructure, lack of monitoring, obsolete equipment and difficult access to qualifications. I found that schools urgently need transformations, such as changes in laboratory facilities for digital learning; transformations that go beyond financial and material resources, and also involve significant actions in teacher training. It is essential that this is continued in today's world in which individuals need to make decisions, take initiatives and seek changes in order to achieve their digital emancipation.

Finally, in order to make sure that this work has achieved its objectives, I have tried to present data that points back to the problematization and objectives raised in the research project. I reiterate that these are reflections from a case study that does not intend to generalize or exhaust the subject, but hopes to serve as a perspective for other related research, with new discussions that broaden the field of study and point to improvements to be implemented in the Brazilian educational system, especially in the digital inclusion policies managed by the MEC.

A lot still needs to be done to achieve a satisfactory level of acceptance and use. With regard to education, some reports by scholars on the subject show that there is a need to evolve even further in terms of thinking, doing, being and being in the world. The big question is: For whom and at whose service are the technologies and their rapid advances available?

I am aware that the subject is wide-ranging, complex and topical, involving education in a broad sense, the technologies of the brave cyber world, pedagogical practices and digital emancipation, which could certainly serve as inspiration for various other studies in the area.

REFERENCES

ADAMS, Telmo. *Education and popular solidarity economy: pedagogical mediations of associated work*. Sao Paulo: Ideias e Letras, 2010.
et al. *Digital technologies and education: for which development? Educaçao Unisinos (Online)*, v. 17, p. 57-65, 2013.
ALMEIDA, Maria Elizabeth Bianconcini de. *Informática e formaçao* de *professores*. Collection Informática Aplicada na Educaçao. Sao Paulo: MEC/SEED/Prolnfo, 1999.
ALMEIDA, Maria Elizabeth. *Prolnfo: Informatics and teacher training. Brasilia:* I *Ministério da Educaçao*, Seed, 2000. 2v.
ALVES, L. R. G. *Novas Cartografias cognitivas - Uma análise do uso das tecnologias intelectuaispor crianças da rede pública em Salvador* - Bahia. Salvador, 1998. 140p. Dissertation (Master's Degree in Education) - Faculty of Education, Federal University of Bahia.
ÂNGELA, Maria Martins; WERLE, Flávia O. Correia (Orgs). *Educational policies: elements for reflection*. Porto Alegre: Redes, 2010.
BENTO, A. (2012, June). *How to prepare (and carry out) a successful defense. Revista JÁ (Academic Association of the University of Madeira),* n° 66, year VII (pp. 42-45). ISSN: 1647-8975.
BRAZIL. Secretariat for Primary Education. *Parámetros Curriculares Nacionais: terceiro e quarto ciclos do Ensino Fundamental: introdução aos Parámetros Curriculares Nacionais.* Brasilia, DF: MEC/SEF, 1998.
BRAZIL, MEC/SEED. *Programa Nacional de Informática na Educaçao PROINFO - Diretrizes*, 1997. Available at http://www.proinfo.mec.gov.br/Acesso on September 23, 2012.
BRAZIL. MINISTRY OF SCIENCE AND TECHNOLOGY. *Information Society in Brazil* - Green Book. Brasilia: 2000.
BRASÍLIA. MINISTRY OF EDUCATION. *One Computer per Student Project (UCA): Working Meeting.* Brasilia-DF, November 07 and 08, 2007.
BRITO, Gláucia da Silva. *Education and new technologies: a rethink.* Curitiba: IBPEX, 2006
CAMBI, Franco. *History of Pedagogy*. Sao Paulo: Unesp, 1999.
COMITÉ GESTOR DA INTERNET NO BRASIL. *Pesquisa Sobre o Uso das Tecnologias de Informaçao e Comunicaçao no Brasil* : *TIC Educaçao 2011*. Sao Paulo: CGI.br, 2012.
Coord. Alexandre F. Barbosa. Translated by Karen Brito.
Available at: <http://www.cetic.br/tic/educacao/2011/index.htm>. Accessed on: July 28, 2013.
CAMPOS, Fernanda A. Coutinho. *Information and communication technologies and teacher training: a study of undergraduate courses at a private university*. Belo Horizonte. 2011. 224f. Dissertation in Education. Federal University of Minas Gerais.
CAPPELLETTI, Isabel Franchi. *Evaluation of the program "One computer per student" (PROUCA): An innovative proposal in public policy.* Revista e-curriculum, Sao Paulo, v.8 n.1 APRIL 2012
CASTELLS, Manuel. *The network society*. Sao Paulo: Paz e Terra, 1999.
CARRETEIRO, Ronald. *Technological innovation: how to ensure business modernity*. Rio de Janeiro: LCT, 2009.
CASTRO, Luciene Sobrinha de (Org.). NTHE 10. A thread of history: Technology in Teresina's municipal schools. Teresina: ADUFPI, 2010
COUTO, Edvaldo. Technology in the classroom is not enough in Brazil.
Available at http://noticias.terra.com.br/educaçào/tecnologia-em-sala-de-aula-nao-e-suficiente-no-brasil-dizpesquisador.html.
Accessed on July 18, 2013.

COSTA, Leonardo Figueiredo. Digital Inclusion: concepts, models and semantics. Paper presented to the NP Tecnologias da Informaçao e da Comunicaçao, of the VI Encontro dos Núcleos de Pesquisa da Intercom, 2006.
DECREE No. 6.300, December 12, 2007. Ministry of Education, Brasilia, 1997. Available at: <http://www.planalto.gov.br/ccivil_03/_Ato20072010/2007/ Decree/D6300.htm>. Accessed on August 17, 2013.
DEMO, Pedro. *Technology in education and learning.* <http://www.edutecnet.com.br/Textos/Alia/MISC/pdemo.htm>. Lecture given on 27/5/2000 at Educador 2000 -- International Education Congress. Accessed on: June 28, 2013.
. *Complexity and Learning - The non-linear dynamics of knowledge.* Sao Paulo: Editora Atlas, 2002.
. *Knowing how to think.* Sao Paulo: Cortez, 2000.
DOWBOR, Ladislau. *Technologies of knowledge* (2001) Available at www.dowbor.org. Accessed on October 15, 2012.
EURÍPEDES AGUIAR SCHOOL. *Institutional information.* Available http://blogdoeuripides.blogspot.com.br/ Accessed September 2012
FREIRE, P. *Pedagogia da autonomia*: saberes necessárias à prática educativa. 7. ed. Sao Paulo, SP: Paze Terra, 1996.
. *Pedagogy of the oppressed.* Rio de Janeiro, RJ: Paz e Terra, 1987.
FREITAS, Adriano Vargas and LEITE, Ligia Silva. With chalk and a *lap top*: from the conception to the integration of public IT policies. Rio de Janeiro, RJ: Wak, 2011.
FREIRE, Wendel (org.). *Tecnologias e educaçao: as mídias naprática docente.* Rio de Janeiro: Wak, 2011.
FREIRE, Fernanda Maria P.; VALENTE, José Armando (Orgs). *Learning for life: computers in the classroom.* Sao Paulo: Cortez, 2001.
FREITAS, Adriano Vargas; LEITE, Ligia Silva. *With chalk and a laptop: from conception to the integration of public policies.* Rio de Janeiro: Wak, 2011.
FERREIRA, M. H. M; FRADE, I. C. A. S. Literacy *and literacy in digital contexts: evaluation assumptions for the HagâQuê software.* In: RIBEIRO, A. E. et al. (eds.). **Language, technology and education**. Sao Paulo: Peirópolis, 2010. p. 15-27.
FRIGOTTO, Gaudêncio. *Technology.* 2009. Available:<http://www.epsjv.fiocruz.br/dicionario/verbetes/tec.html>. Accessed on: July 18, 2013.
GADOTTI, Moacir. *Pedagogy of Praxis.* Sao Paulo: Cortez, 2010.
GIL, Antonio Carlos. *How to prepare research projects.* Sao Paulo: Atlas, 2010.
IBGE, Brazilian Institute of Geography and Statistics Available at: http://www.ibge.gov.br/home/mapa_site/mapa_site.php#populacao. Accessed on August 20, 2013.
LEMOS, André; LEVY, Pierre. *The future of the internet.* Sao Paulo: Paulus, 2012.
LEVY, Pierre. *Cyberculture.* Sao Paulo: Editora 34, 1999.
LEVY, P. *The technologies of intelligence.* Sao Paulo: Ed. 34, 2004.
LIMA, Patrícia Rosa Traple. *New information and communication technologies and teacher training in degree courses in the state of Santa Catarina.* 2001. 83f. Dissertation in Computer Science. Federal University of Santa Catarina.
LIMA, Frederico O. *A sociedade digital: impacto da tecnologia na sociedade, cultura, na educação e nas organizações.*Rio de Janeiro: Qualitymark Editora,2000.
LOPES, Daniel de Queiroz; SCHLEMMER, Eliane. *Digital culture in schools: beyond the question of access to digital technologies.* Nov. 2011. Available at: http://www.simp0sio2011.abciber.org/anais/trabalhos/artigos/Eixo%201/9E1/365-591-1- RV.pdf. Accessed on October 20, 2012.
LOPES, Daniel de Queiroz. *Playing with robots: designing problems and inventing whys.* Santa Cruz do Sul, RS: Edunifc, 2010.

MARQUES, António Carlos Conceiçao. *The one computer per student project (UCA): Reactions in schools, teachers,* students and *institutions.* Curitiba, 2009. 98f. Dissertation.
Master's Degree in Education - Federal University of Paraná (UFPR).
MASETTO, M. T. Pedagogical mediation and the use of technology. In: MORAN, J. M.;
MASETTO, M. T. BEHRENS, M. A. *Novas tecnologias e mediaçãoopedagógica.* Campinas: Papirus, 2000 (p. 133-173).
MEIRELES, Alcides José da Costa. *Using interactive whiteboards in education: an experience in physical chemistry with advantages and "resistances".* 140f. in 2006. Dissertation in Education. University of Porto - Portugal.
MINISTRY OF DEVELOPMENT, INDUSTRY AND COMMERCE. *UCA Project - One Computer per Student.* Available at www.mdic.gov.br Accessed on: October 27, 2012.
MINISTRY OF PLANNING, BUDGET AND MANAGEMENT. Management Secretariat. Public Management for a Brazil for All: a management plan for the Lula government. Brasilia: MP, SEGES, 2003.
MORAN, José Manuel; MASETTO, Marcos T.; BEHRENS, Marilda Aparecida. *New technologies and educational mediation.* Sao Paulo: Papirus, 2001.
MORAES, Maria Cándida. *Educational informatics in Brazil: a history lived, some lessons learned.* Available at:
<http://www.edutec.net/Textos/Alia/MISC/edmcand1.htm. Accessed September 22, 2012.
OLIVEIRA, Maria Rita Neto Sales. *Interactive technologies and education.Educaçao em Debate.* Fortaleza: Year 21, n. 37, p. 150-156, 1999.
From the myth of technology to the technological paradigm; technological mediation in didactic-pedagogical practices. Revista Brasileira de educaçao, n° 18, p.101107, Sep.-Dec.2001
Available:
http://www.anped.org.br/rbe/rbedigital/rbde18/rbde18_10_maria_rita_neto_sales_oliveira.pdf.
Accessed on: October 16, 2012
OLIVEIRA, Marta Kohl. *Vygotsky: Learning and development. A socio-historical process.* Sao Paulo: Scipione, 1993.
PACHECO, Márcia Arantes Buiatti. *Digital education: a perspective of inclusion in everyday school life.* 2011. 172 f. Dissertaçao(Mestrado)-Universidade Federal de Uberlândia, Uberlândia, 2011. Available at http://hdl.handle.net/123456789/969. Accessed July 18, 2013.
PAIVA, Vera Lúcia Menezes de Oliveira e. *The use of technology in foreign language teaching: a brief historical retrospective.*Belo Horizonte: UFMG/CNPq/FAPEMIG.
PERRENOUD, Philipe. *Differentiated pedagogy: from intentions to action.* Porto Alegre: Artmed, 2000.
PROINFO: *Informática e formaçao deprofessores/Secretaria.Educaçao a Distância.* Brasilia: Ministry of Education, SEED, 2000.
REGO, Teresa Cristina. *Vigotsky: a cultural-historical perspective on education.* Petrópolis, RJ: Vozes, 2011.
ROSINI, Alessandro Marco. *The use of computer technology in education. A reflection on teaching children.* Available at www.ipv.pt/millenium/millenium27/15.htm. Accessed on: September 13, 2013
SANTOS, Milton. *The Nature of Space: Technique and Time, Reason and Emotion.* 4. ed. Sao Paulo: Editora da Universidade de Sao Paulo, 2006.
SANTOS, Milton. *Technique Space Time: globalization and the technical-scientific*

informational environment. 2nd ed. Sao Paulo: Hucitec, 1994.
SILVA, Antonio Mendes da. *The three pillars of digital inclusion.* Available at www.espacoacademico.com.br. Accessed October 2012
SILVA, Maria Reginalda Soares. *Continuing teacher training: knowledge and reflexivity in pedagogical practice.* 2010. 223f. Dissertation in Education. Federal University of Piaui (UFPI).
SCHLEMMER, Eliane. *Politicas e práticaspedagógicas na formaçao de professores a distância: por uma emancipaçao digital cidada.*
SCHWARTZ, G. *Educating for digital emancipation.* Available at: <http://www.reescrevendoaeducacao.com.br/>. Accessed in October 2012
TEDESCO, Juan Carlos (Org.). *Education and New Technologies: hope or uncertainty?* Sao Paulo: Cortez, 2004.
VALENTE, José Armando. *Informática na educaçao.*Disponível em http://www.nte-jgs.rctsc.br/valente.htm
VALENTE, José Armando. An *analytical view of information technology in education in Brazil: the question of teacher training.* Available at http://www.pucpr.br/eventos/educere/educere2009/anais/pdf/1919_1044.pdf Accessed on: 18 Nov. 2013
. *How many clicks does a lesson take? Reflections on digital literacy - the new challenge for schools.* Available at: <http://www.escoladositio.com.br/ce/pal_ant.htm#ant04>. Accessed on: October 10, 2012.
VASCONCELOS, Celso dos S. *Construgáo do Conhecimento em sala de aula.* Sao Paulo: Libertad, 2005.
VIEIRA, Sonia. *How to design questionnaires.* Sao Paulo: Atlas, 2009.
WERTSCH, J.V (org.) *Vygotsky and the social formation of the mind.* Barcelona: Paidós, 1988.

APPENDIX

APPENDIX A - Invitation letter - Participation in the survey
University of Vale do Rio dos Sinos (UNISINOS-RS)
Postgraduate Program in Education
MINTER: IFPI/UNISINOS
Research Line: Education, development and technologies
Master's student: Elizabete Rodrigues Sales Advisor: Prof. Dr. Telmo Adams

Dear Teacher,

I, Elizabete Rodrigues Sales, a permanent employee of the Federal Institute of Education, Science and Technology of Piauí (IFPI), located in the city of Teresina (PI), teacher of the Technological Axis of Management and Services, request your collaboration in order to participate in a survey of which I am part and which corresponds to a stage of the Master's Degree in Education.

The title of the research I am now undertaking is Digital Technologies: Their Place in Pedagogical Practices in a Municipal Public School in Piauí.

In this scientific work, my general objective is to analyze the configuration of the educational reality with regard to the adequacy of spaces, access to the use of computers, the Internet and other digital technologies in the pedagogical practices of teachers, after the implementation of the National Program for Informatics in Education (Prolnfo).

Thank you in advance for your cooperation.

Teresina/PI, August 2013

Elizabete Rodrigues Sales

Master's student

APPENDIX B - Questionnaire 1 - First stage - Selection of research participants

I Identification data

1) Name:____________________________________
2) Pseudonym (to appear in the search): ____________________
3) Contact telephone number: __________________________
4) Age:______________________________________
5) E-mail:____________________________________
6) Sex: ( ) Female
( ) Male
7) Area taught:Series:
8) Subject: ____________________________________
9) Length of service in the position or function at the school
10) Functional status: ( ) Permanent
( ) Not effective

II Information on digital technological resources and pedagogical practice

11) Do you usually use technological resources, such as computers and the *internet,* in teaching and learning activities? How often and in what way? Give an example.

____

____

____

____

____

12) In your opinion, is it possible to enhance school work using computers and the *Internet?* Explain.

____

____

____

____

____

13) Have you ever developed or participated in an interdisciplinary pedagogical project using and integrating digital technologies?

12) Have you received or do you receive any training in the use of these technological resources in your school?
( ) Yes ( ) No

____

____

____

____

____

14) If yes. How often? If you wish, please give your opinion.

____

____

____

____

____

APPENDIX C - Informed Consent Form
Title of the study:Digital Technologies: Their place in pedagogical practices in a municipal public school in Piauí
Researcher responsible: Elizabete Rodrigues Sales (PPGEDU/UNISINOS)
Advisor: Prof. Dr. Telmo Adams (PPGEDU/UNISINOS)
Study period: February 2012 to February 2014

Invitation to participate in the study

You are being invited to take part in a master's research project under the responsibility of Elizabete Rodrigues Sales and the supervision of Prof. Dr. Telmo Adams of the Digital Education Research Group of the Postgraduate Program in Education at the University of Vale do Rio dos Sinos (GPe-dU/PPGEDU/UNISINOS).

In order to decide whether or not to take part in this research, you need to know about the aims of this study. This Informed Consent Form provides detailed information about the research, which will be presented and discussed with you.

After receiving information about this study, you will be asked to sign this informed consent form if you agree to take part. Ask the research coordinator or someone from your team to explain any questions you may have before you sign this informed consent form.

What is the aim of this study?

The research addresses the issue of digital technology (DT), involving the computer, information and communication technology (ICT) and the *internet*, configured as mediating resources in the pedagogical practices of teachers in the teaching process, from there, its implications and contributions in the school in contemporary times. It aims to understand the meanings of pedagogical mediation and digital emancipation in the teaching process, offering subsidies for understanding the role of digital technologies, the new profile of the school and the teacher in education.

This is a qualitative research project, which is a case study and seeks to obtain from the participants, in the field of subjectivity, their understandings and meanings about the object of study. To this end, we wanted to carry out this research in a municipal public school in Piauí, whose teachers have at their disposal a computer laboratory acquired through the National Program for Information Technology in Education (ProInfo).

Bearing in mind the assumption that digital emancipation is associated with training, technological appropriation, as well as competence and didactic-pedagogical skills, the theoretical basis for this dissertation includes the meanings of pedagogical mediation, digital technologies and digital emancipation. These are also used as categories for the research, which aims to understand the conceptualization given by teachers regarding the use of digital technologies in teaching practices.

What are my responsibilities if I take part in this study?

You will take part as an interviewee; at first, I intend to conduct an individual interview which will be scheduled with you in advance. The interview will be semi-structured, i.e. there will be some questions formulated in advance by the researcher, but at the time of the dialogue, the speech will be freely organized by you, always focusing on the research topic.

To ensure the utmost fidelity to your speech, the entire interview will be recorded and then transcribed. The entire audio will be made available to you as well as the transcript.

And what about the confidentiality of the information collected by the researcher?

The researcher involved in the project undertakes to keep the identity of the research participants confidential, as well as any others who may be mentioned during the process, including institutions of any kind. No names or any other data that would allow them to be identified will be disclosed. Participants will be referred to by pseudonyms defined by them in advance. All the information collected will be organized in digital databases with restricted access to the researchers, and will be stored for up to 10 years (from the end date of this research) and then deleted. You can access your data at any time by asking the coordinator or researcher responsible for the study.

Who else will take part in this study?

Participants in this study included the school manager and previously selected teachers who work with digital technologies and use the PROINFO computer lab.

Can I withdraw from this study?

You can withdraw from participating in this research at any time, without any harm to you. To do so, simply notify the researcher by telephone or e-mail.

Will I receive payment for taking part in this study?

No. Participants will not receive any payment for taking part in this research.

Will there be any costs involved?

No. You will not incur any additional costs by participating in this survey.

If I have any questions or problems, who should I contact?

If you need any additional information, have any questions, suggestions, complaints, or wish to inform us that you no longer wish to take part in the research, you can contact the person responsible for this research, Elizabete Rodrigues Sales, directly by telephone (86)9956-1223/88213435 or by e-mail criativa_beth@yahoo.com.br, or the research supervisor, Prof. Dr. Telmo Adams, by telephone (51)3590-8241 or by e-mail <telmoa@unisinos.br>.

I therefore certify the following:

- I have read the information above and understand that the study involves research. I am aware of the purpose of the study.
- I had the opportunity to clarify my doubts. All my questions regarding this study were answered satisfactorily.
- I understand that I am free to withdraw from this study at any time.

I agree to participate in this study and understand that I will receive a signed copy of this Informed Consent Form.

Participant's name (print) Date

SIGNATURE OF THE RESEARCHER RESPONSIBLE:

Elizabete Rodrigues Sales

Researcher

Researcher's signatureDate

SUPERVISOR'S SIGNATURE:

Telmo Adams

Advisor

Advisor's signatureDate

APPENDIX D - Interview script - Teachers

I) THEMATIC AXIS 1 - Experience with digital technologies

1) In your opinion, how do digital technologies enable experience, innovation and research in the school environment? Is there a project at your school that covers these areas?
2) Do you think that the use of the Computer Lab, acquired by PROINFO, is changing the way teachers carry out their pedagogical work or not?
3) How has this practice improved? Can you give an example?
4) What are the prospects for change in your teaching practice with regard to the democratization of global information through the Internet?
5) What are the biggest difficulties encountered in your school when it comes to including digital technologies in your day-to-day teaching?
6) In your opinion, what is missing for digital inclusion to really happen in your school? What do teachers need in order to include digital technologies in their teaching practice?
7) Do you use any resources available on the Internet (social media) as a way of interacting with your students? Which ones and how does this interaction work? Does it enable the construction of knowledge?
8) What are the main needs that you think are essential for changes in your pedagogical performance, using digital technologies as a reference?
9) Do you use computers and the Internet in your teaching activities outside the classroom? What are these activities?
10) In your opinion, what does a teacher need to develop a good lesson?

II) THEMATIC AXIS 2 - CAPACITY BUILDING - Influence on teaching practice: changes and evolution

1) Have you ever attended a training course in the pedagogical use of computers? Was it your initiative or that of the Municipal Department of Education (SEMEC)? SEMEC being the case - Are the courses ongoing and do they meet teachers' needs in terms of preparing them to use computers and the Internet in their teaching practices?
2) Are you currently taking any computer training courses? Are they only technical courses or do they offer any pedagogical software?

III) THEMATIC AXIS 3 - PROINFO - Encouraging and influencing teaching practices in the use of DM

1) In your opinion, does the PROINFO Program contribute to the digital inclusion of teachers? Please explain.
2) Does the computer lab provided by PROINFO operate in favorable conditions for the teacher to develop their lesson as planned?

3) What are the main activities you carry out in the Computer Lab? And what advantages and benefits have you seen from this practice?
4) In your opinion, does PROINFO enable or promote changes in traditional teaching practices? If there are changes, do these new practices enhance or contemplate the stimulation of learning?
5) If you wish, give your opinion on the program under analysis: PROINFO in your school.

APPENDIX E- Interview script - Coordination of the NTHE

1) When was ProInfo implemented in the municipality?
2) How is the selection of schools to receive ProInfo laboratories made?
3) Do you have a partnership for this process?
4) Are the teachers in the municipality's schools receptive to these courses offered by the NTHE?
5) To meet the demands you mentioned, does the center wait for a request from the school? Is it the school's initiative?
6) What is the renewal period for this pact, the contract for the company that supplies the equipment to the MEC?
7) What other support does the center offer schools?
8) In the universe of teachers served by the NTHE, what percentage of them already use the laboratory and technologies? Do you have an estimate?
9) Is the internet system the responsibility of the NTHE?
10) The laboratory at the school where I am doing my research is also available for the "Mais Educaçâo" Program. In this case, is the NTHE also available to support the program?
11) In your opinion, how does ProInfo work today?
12) What is the situation of the NTHE today and what is your opinion of this MEC initiative with ProInfo?
13) Do you think that, in this situation, the greatest interest lies with the teacher themselves, who is committed to finding the time to invest in this training?
14) The teachers at the school I'm working with in my research were pleased with this initiative by the MEC and some even said that it was through ProInfo that they became familiar with the resources and started using digital technologies and incorporating them into their pedagogical projects and lesson plans. Did you already have this information?
15) With regard to the Internet, do all schools have it?
16) What progress has the program made since its implementation?
17) Does the NTHE have the autonomy to set up workshops and courses it deems necessary?
18) Do you have any more information you'd like to add?

ANNEX A - Ordinance No. 522 of April 9, 1997

MINISTRY OF EDUCATION AND SPORT
MINISTER'S OFFICE
Ordinance No. 522 of April 9, 1997

THE **MINISTER OF STATE FOR EDUCATION AND SPORT**, in the use of his legal attributions, resolves to

Art. 1 - The National Program for Information Technology in Education - ProInfo is hereby created, with the aim of disseminating the pedagogical use of information technology and telecommunications in public elementary and high schools belonging to the state and municipal networks.

Sole Paragraph. ProInfo actions will be developed under the responsibility of the Ministry's Distance Education Secretariat, in conjunction with the education secretariats of the Federal District, States and Municipalities.

Art. 2 The statistical data needed for planning and allocating ProInfo resources, including enrollment estimates, will be based on the school census conducted annually by the Ministry of Education and Sports and published in the Official Gazette of the Union.

Art. 3 The Secretary of Distance Education will issue norms and guidelines, establish criteria and operationalization and adopt the other measures necessary for the execution of the program referred to in this Ordinance.

Art. 4 This Ordinance shall enter into force on the date of its publication.

PAULO RENATO SOUZA

Minister of Education and Sport

ANNEX B - DECREE NO. 6.300, OF DECEMBER 12, 2007

Presidency of the Republic

Civil House

Deputy Head of Legal Affairs

DECREE NO. 6.300, OF DECEMBER 12, 2007.

Provides for the National Educational Technology Program -ProInfo.

THE PRESIDENT OF THE REPUBLIC, in the exercise of the powers conferred on him by article 84, items IV and VI, paragraph "a", of the Constitution, and in view of the provisions of Law No. 10.172, of January 9, 2001,

DECREE:

Art. 1 The National Educational Technology Program - ProInfo, executed within the scope of the Ministry of Education, will promote the pedagogical use of information and communication technologies in public basic education networks.

Sole paragraph. The objectives of ProInfo are:

I - to promote the pedagogical use of information and communication technologies in basic education schools in urban and rural public education networks;

II - to encourage the improvement of the teaching and learning process through the use of information and communication technologies;

III - promote the training of educational agents involved in the Program's actions;

IV - contribute to digital inclusion by expanding access to computers, connections to the World Wide Web and other digital technologies, benefiting the school community and the population close to the schools;

V - contribute to preparing young people and adults for the job market through the use of information and communication technologies; and

VI - to promote the national production of digital educational content.

Art. 2 ProInfo will fulfill its aims and objectives in a system of collaboration between the Union, the States, the Federal District and the Municipalities, through adhesion.

Art. 3 The Ministry of Education is responsible for:

I - set up technological environments equipped with computers and digital resources in the beneficiary schools;

II - promote, in partnership with the States, Federal District and Municipalities, a training program for the educational agents involved and the connection of technological environments to the World Wide Web; and

III - provide educational content, solutions and information systems.

Art. 4 The States, the Federal District and the Municipalities that join ProInfo are responsible for:

I - provide the necessary infrastructure for the proper functioning of the Program's technological environments;

II - enable and encourage the training of teachers and other educational agents in the pedagogical use of information and communication technologies;

III - ensure human resources and the necessary conditions for the work of support teams for the development and monitoring of training actions in schools;

IV - ensure technical support and maintenance of the equipment in the Program's technological environment, once the contracted supplier company's warranty period has expired.

Sole paragraph. Education networks must include the use of information and communication technologies in the political-pedagogical projects of the schools benefiting from ProInfo.

Art. 5 ProInfo's expenses shall be borne by the budget appropriations allocated annually to the Ministry of Education and the National Education Development Fund - FNDE, and the Executive Branch shall make the selection of courses and programs compatible with the existing budget appropriations, observing the movement and commitment and payment limits of the budget and financial programming defined by the Ministry of Planning, Budget and Management.

Art. 6 The Ministry of Education will coordinate the implementation of technological environments, monitor and evaluate ProInfo.

Art. 7 An Act of the Minister of State for Education will establish the operational rules and adopt the other measures necessary for the implementation of ProInfo.

Art. 8 This Decree enters into force on the date of its publication.

Brasilia, December 12, 2007; 186th of Independence and 119th of the Republic.

LUIZ INÁCIO LULA DA SILVA

Fernando Haddad

This text does not replace the one published in the DOU of 13.12.2007

The basic understanding of the natural and social environment, the political system, technology, the arts and the values on which society is founded.

Printed by Books on Demand GmbH, Norderstedt / Germany